Underutilized Fruits

A Colour Handbook

The Author

Dr. P. C. Tripathi is working as Head (IC), Division of Plant Genetic Resources, IIHR, Bangalore. He worked as scientist, Senior scientist, and Principal scientist, Section In-charge, Head, Regional station and Krishi Vigyan Kendra in different ICAR institutes and SAUs. He has been associated with more than 25 research projects including 10 externally funded projects. He released four varieties of fruits and identified 10 promising lines of underutilized fruits and develops 6 prototypes of equipments and 10 production and post- harvest technologies of mango, onion and garlic etc. Last ten years, he is working on underutilized and wild fruits. He published more than 100 research papers, 129 research abstracts, 49 popular articles, 40 technical bulletins, 5 books, 30 book chapters etc. He is associated with several professional societies and received several awards.

Underutilized Fruits

A Colour Handbook

P. C. Tripathi

ICAR-Indian Institute
of Horticultural Research,
Bangalore

2022

Daya Publishing House®

A Division of

Astral International Pvt. Ltd.

New Delhi – 110 002

ISBN: 978-93-5461-314-2

Published by : **Daya Publishing House®**
A Division of
Astral International Pvt. Ltd.
– ISO 9001:2015 Certified Company –
4736/23, Ansari Road, Darya Ganj,
New DelhiA110 002
Ph. 011A43549197, 23278134
EAmail: info@astralint.com
Website: www.astralint.com

Digitally Printed at : **Replika Press Pvt. Ltd.**

Preface

Underutilized fruit crops refer to those fruits which may be high in value but that are not widely grown. An exact definition of underutilized fruit crops is perhaps difficult. In a general sense, these fruits are consumable by in relatively less quantity may be due to less palatable or less availability than other fruits. These fruits may have lesser demand in the market or grown in a limited extent. Other terms for these fruits are less-known, less appealing, less-exploited fruits, underutilized, stray fruits, wild fruits etc. However, any sharp line of distinction between the major and underutilized fruits is difficult. A fruits which is major fruit in one region or country may be minor in other region or country. The underutilized fruits may be categorized in two groups i.e. indigenous and exotic. India has a rich and varied heritage of biodiversity, encompassing a wide spectrum of habitats from tropical rainforests to alpine vegetation and from temperate forests to coastal wetlands. Out of 18 biodiversity hot spots identified in the world, four hotspots, i.e. Western Ghats, Eastern Himalaya, Western Himalaya, and Nicobar islands are in India. Several fruit plant species have originated in Indian subcontinent. India is centre of origin of jack fruit, bael, aonla, ber, jamun, mahua, phalsa, Lasoda, karonda, wood apple, bilimbi, Garcinia, and several other wild fruits. Some of these fruits have gained popularity in past few decade but majority are still not grown in large area. Apart from indigenous fruits , several fruits were introduced in India from South America , Africa, South East Asian countries during last four centuries. Some of adopted to climatic conditions of India and become major fruits. But some remain confined to specific locations or backyard gardens due to several reasons. Several minor fruits such as Rambutan, mangosteen, longan, avocado, water apple, hog plum, macadamia nut, kiwifruit, longsat, durian, passion fruit, dragon fruit, pulasan, cararmbola, etc. were introduced during last few centuries .These fruits are rich in certain nutrients and slowly these are making place in our food baskets.

The precise information on these fruits is compiled in this book to provide summarised information to the student and general public.

P. C. Tripathi

Contents

Introduction

Underutilized fruit crops refer to those fruits which may be high in value but that are not widely grown. An exact definition of underutilized fruit crops is perhaps difficult. In a general sense, these fruits are consumable by in relatively less quantity may be due to less palatable or less availability than other fruits. These fruit s may have lesser demand in the market or grown in a limited extent. Other terms for these fruits are less-known, less appealing, less-exploited fruits, minor fruits, stray fruits, wild fruits etc. However, any sharp line of distinction between the major and minor fruits is difficult. A fruits which is major fruits in one region or country may be minor in other region or country. For example Rambutan in major fruits crops in Indonesia, Malaysia but in minor crops in India. Similarly mango is the most important fruit in India while it is regarded as a minor fruit crop only in some western countries. If area and production are considered to be the criteria to call a fruit crop as major or minor, difference may be observed even in the same country as well.The climatic conditions are important for cultivation of a particularly crop. This makes some fruits to grow extensively over large area in a region unlike other parts in a country. It is estimated that there are 8.7 million species of the world's biota. India contributes significantly to the biodiversity of the world by accounting 7.31 % of the global plant diversity from 2.4% of the world's area. It is one of main centre of crop diversity and the homeland of 167 cultivated species and 320 wild relatives of crop plants. Several fruit plant species have originated in Indian subcontinent. India is centre of origin of jack fruit, bael, aonla, ber, khejri, jamun, tamarind, mahua, phalsa, Lasoda, karonda, wood apple, pilu, bilimbi, Garcinia, and several other wild fruits . Several minor fruits such as Rambutan, mangosteen, longan, avocado, water apple, hog plum, macadamia nut, kiwifruit, longsat, durian, passion fruit, dragon fruit, pulasan, carambola, etc. were introduced during last few centuries and several are naturalized in Indian conditions . Little research works have been carried out India on underutilized fruits because of their limed or no commercial cultivation. There are very few released varieties of these fruits. The standardized package of practices of these fruits is not available. The grower used the practices available in other fruits for their cultivation. Several underutilized fruits are rich in vitamin, minerals, nutrients and have medicinal potential. There is growing interest in these crops to contribute in novel ways to human health and nutrition. This interest can be exploited to develop markets for non-staple crops from which poor communities can benefit, providing incentives for farmers to plant these crops. These fruits are facing problems of loss of germplasm material, traditional knowledge. The lack of knowledge their uses, limited income generation, limited market, and demand, lack of research and development activities are other challenges

P. C. Tripathi

1 AVOCADO

Avocado	Native of Central- South American countries, Avocado is the most nutritive among fruits. Pulp is rich in proteins (up to 4%) and fat (up to 30%) and low in carbohydrates, highest energy value (245 cal/100 g), rich in copper and iron
Botanical name	*Persia americana* Mill
Family	Lauraceae
Commercial varieties	Fuerte, Bacon, Zutano, Hass, Booth 7, Booth 8, Sharwil, Ettinger, TKD-1, Arka Supreme
Economic part	Fruit , seed
Propagation techniques	Cleft grafting
Nursery period	12-18 months
Grafting time	March & September
Spacing	6 m x 6 m
Climate and soil requirements	Tropical or sub tropical climate with good rainfall, Intolerant to saline conditions, optimum pH range is from 5 to 7. Extremely sensitive to water-logging.
Leaf	Simple
Growth form	Upright, divergent , drooping
Flowering period	Oct - Nov., Feb- March
Pollination	Cross pollination

Flowering tree of avocado

Flowering panicle of avocado

Avocado fruiting branch

Pollinating agents	Bees	 Mature fruits of avocado
Economic yield	After 8 years	
Yield/tree	40-60 kg/tree	 Anthracnose disease of avocado
First harvest	4- 5 years	
Harvesting method	Fruit with stalk	
Harvesting period	6-8 month after fruit set	
Physiological disorders	Nil	
Pests	Mites, mealy bug, scale	
Diseases	Anthracnose, Phytophthora root rot, leaf spot, Stem rot	
Shelf life	One week	
Shelf life in cold storage	Up to four weeks at 5.5 to 8°C.	
Post harvest products	Mainly used fresh, Also used in sandwich filling, salads, ice creams, milk shake, pulp preserved by freezing, powder, paste	

Nutritional value of Avocado Fruit (per 100g of edible portion)

Nutrients	Quantity	Nutrients	Quantity
Protein	1.7 g	Calcium	10.00 mg
Fat	26.4 g	Chlorine	11.00 mg
Total carbohydrates	5.1 g	Copper	0.45 mg
Crude fibre	1.8 g	Iron	0.60 mg
Vitamin A as carotene	0.17 mg	Magnesium	35.00 mg
Ascorbic acid	16.00 mg	Manganese	4.21 mg
Niacin	1.10 mg	Phosphorus	38.00 mg
Riboflavin	0.13 mg	Sodium	368.00 mg
Thiamine	0.06 mg	Sulphur	28.50 mg

2 BADAVAPULI

Badavapulli	Grows mostly wild in South India and southern parts of the Central India, a type of gajnimba
Botanical name	*Citrus pennivesiculata* (Lush). Tanaka
Family	Rutaceae
Commercial varieties/lines	-
Economic part	Fruit
Propagation techniques	Seeds, patch or shield Budding
Nursery period	9-15 months
Multiplication time	July -September
Spacing	6 m x 6 m
Climate and soil requirements	Subtropical, well drained sandy loam with Ph 5.5 to 7.5
Leaf	Simple
Growth form	Perennial, semi vigorous
Flowering period	Feb - March, Sept.
Pollination	Cross pollination
Pollinating agents	Honey bees
Economic yield	After 10 years
Yield	full grown tree produces from 40-100 fruits
First harvest	5-8 years

Badwapuli fruits

Fruiting branch

Harvesting method	Fruit manually with or with out stalk when colour changes to yellow
Harvesting period	6-7 months after fruit set
Physiological disorders	nil
Pests	Soft green scale, , mites, aphids , psylla, fruit fly
Diseases	Phytophthora rot, Powdery mildew, Citrus Greening disease, Citrus Tristeza virus
Shelf life	Two week
Shelf life in cold storage	-
Post harvest products	Pickle

Mature fruits of badwapuli

Greening disease

Nutritional value of Badavapuli fruits per 100 g edible portion

Nutrients	Quantity	Nutrients	Quantity
Moisture (%)	86.2	TSS (Brix)	8.1
Carbohydrate(%)	7.3	Acidity (%)	2.75
Protein (%)	0.91	Juice (%)	16.46
Fat (%)	0.84	Vitamin C	13.4 mg

3 BAEL

Bael (Wood apple)	Woody tree, native to India, grown in India, Sri Lanka, Pakistan, Bangladesh, Myanmar, Thailand and most of south-eastern Asian countries.
Botanical name	*Aegle marmelos* (L.) Correa
Family	Rutaceae
Commercial varieties/lines	CISH B-1, CISH B-2, NB 5, NB 6, Pant Shivani, Pant Aparna
Economic part	Leaf, Fruit, Seed, Root
Propagation techniques	Seeds, recalcitrant and can not be stored for longer periods , patch or shield Budding, Air layering
Nursery period	9-15 months
Multiplication time	July -September
Spacing	8-10 m x 8-10m
Climate and soil requirements	Any type of soil such as sandy, clay, water logged, unirrigated, acidic or alkaline in the pH range of 5–10.

Young Bael Tree

Fruit Variability in Bael

Leaf	trifoliate, alternate, each leaflet 5-14 x 2–6 cm, ovate with tapering or pointed tip and rounded base, untoothed or with shallow rounded teeth. Young leaves are pale green or pinkish, finely hairy , mature leaves dark green and completely smooth. leaflets have long stalk	 Ripe fruit of Bael
Growth form	Perennial, vigorous	 Bael shabet
Flowering period	April -June	
Pollination	Cross pollination	
Pollinating agents	Honey bees, Carpenter bee	
Economic yield	After 10 years	
Yield	full grown tree produces from 40-100 fruits	
First harvest	5-8 years	
Harvesting method	Fruit manually with or with out stalk when colour changes to yellowish green	
Harvesting period	10-11 months after fruit set	 Fruit cracking in Bael
Physiological disorders	Fruit cracking	
Pests	Lemon butterfly *(Papiltodemoleus)*, Bael fruit fly *(Bactricera zonata)*	
Diseases	Bacterial shot hole, fruit canker and gummosis	
Shelf life	2-3 weeks	
Shelf life in cold storage	-	
Post harvest products	Squash, preserve (murabba)	

Nutritional value of Bael fruits (per 100 g pulp)

Nutrients	Quantity	Nutrients	Quantity
Moisture	61.5 g	Thiamine	0.13 mg
Carbohydrate	32.1 g	Riboflavin	1.19 mg
Protein	1.8 g	Niacin	1.1 mg
Fat	0.3 g	Calcium	85 mg
Fibre	2.9 g	Potassium	600 mg
Vitamin A (carotene)	55 mg	Energy	137 kcal

4 BARBADOS CHERRY

Barbados cherry, West Indian cherry	Native to South America and Central America, grown in tropical regions including India, can bears upto -1 °C.
Botanical name	*Malpighia glabra* Millsp
Family	Malpighiaceae
Commercial varieties/lines	Acerola, Florida sweet
Economic part	Fruits
Propagation techniques	Seed, cutting
Nursery period	12-15 months
Multiplication time	July -September
Spacing	4-6 m x 4-6 m
Climate and soil requirements	Tropical climate, Sandy, Sandy Loam, , Well-drained.
Leaf	Simple, shiny green
Growth form	Perennial, bushy small tree
Flowering period	April -June
Pollination	Cross pollination
Pollinating agents	Honey bees
Economic yield	After 8 years
Yield	20 - 40 kg of fruits/tree
First harvest	After 3 years

A grown up plant of Barbados cherry

Flowers of Barbados cherry

Fruiting in Barbados cherry

Harvesting method	Manually or shaking tree /branches	
Harvesting period	60-70 days after fruit set	 Unripe, mature & ripe fruits of Barbados cherry
Physiological disorders	-	
Pests	Fruit fly	
Diseases	-	
Shelf life	1-2 days	
Shelf life in cold storage	3 days at 7-8 ° C	
Post harvest products	Syrup, wine	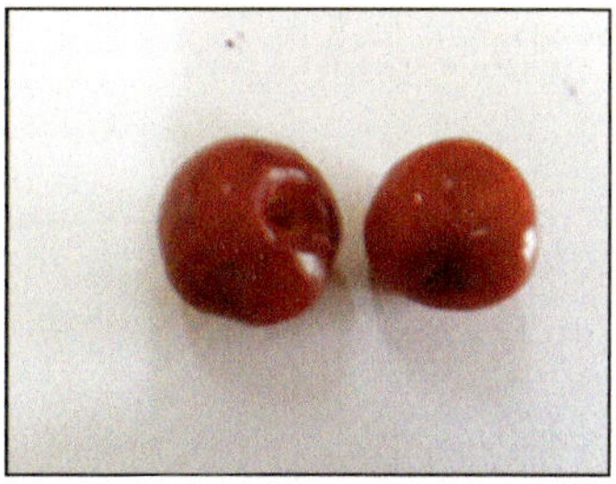 Fruit Fly infested fruits of Barbados cherry

Nutritional value of Barbados cherry fruits (per 100 g edible portion)

Nutrients	Quantity	Nutrients	Quantity
Moisture	92.0%	Vitamin B_3	0.4 mg
Carbohydrate	7.69 g	Vitamin C	1677.6 mg
Protein	0.4 g	Calcium	12 mg
Fat	0.3 g	Phosphorus	11 mg
Fibre	1.1 g	Potassium	146 mg
Vitamin A (carotene)	38 ug	Iron	0.02 mg
Energy	32 kcal	Magnesium	18 mg
Vitamin B_1	0.02 mg	Sodium	7 mg
Vitamin B_2	0.06 mg		

5 BILIMBI

Bilimbi	Native of the South East Asia , evergreen, medium height tree, cultivated throughout Indonesia, Philippines, Sri Lanka Myanmar, Thailand, India.	 Grown up tree of bilimbi
Botanical name	*Averrhoa bilimbi* L.	
Family	Oxalidaceae	
Commercial varieties/lines	-	
Economic part	Fruits	
Propagation techniques	Seeds , T- budding and air-layering	
Nursery period	12-18 months	 Fruiting in bilimbi
Multiplication time	July -September	
Spacing	6 m x 6 m	
Climate and soil requirements	Tropical or subtropical climate, moderate precipitation, more sensitive to cold than the carambola, deep, fertile, well-drained soil	
Leaf	Evergreen, compound, alternate, imparipirmate mainly clustered at the branch tips, leaflets, ovate or oblong, with rounded base and pointed tip	

Growth form	Perennial, vigorous
Flowering period	Feb- March and July - August
Pollination	Cross pollination
Pollinating agents	Honey bees, Carpenter bee
Economic yield	After 6 years
Yield	100-200 fruit/tree
First harvest	4-5 years
Harvesting method	Fruit manually with or without stalk
Harvesting period	Dec -Jan, April-May
Physiological disorders	Fruit drop
Pests	Nil
Diseases	Fruit rot
Shelf life	2-3 days
Shelf life in cold storage	-
Post harvest products	Chutney, pickle

Mature fruits of Bilimabi

Fruit rot in Bilimbi

Nutritional value of Bilimbi fruits (Per 100 g of edible portion)

Nutrients	Quantity	Nutrients	Quantity
Calories	19 kcal	Iron	1.01 mg
Moisture	94.2 g	Phosphorus	11.1 mg
Protein	0.61 g	Carotene	0.035 mg
Fat	0.0 g	Thiamine	0.010 mg
Fiber	0.6 g	Riboflavin	0.026 mg
Ash	0.31-0.40 g	Niacin	0.302 mg
Calcium	3.4 mg	Ascorbic Acid	15.5 mg

6 BRAZILIAN GUAVA

Brazilian Guava	A slow growing shrub or small tree between 2 and 4m . more attractive in foliage and fruit than the common guava, native to the Mexico to Argentina, It is naturalized in parts of India
Botanical name	*Psidium guineense* Sw
Family	Myrtaceae
Commercial varieties/lines	-
Economic part	Fruits
Propagation techniques	Seed, layering,
Nursery period	12 months
Multiplication time	July - Sept
Spacing	3 x 3 m
Climate and soil requirements	Tropical and subtropical climate , does well in moist, fertile soils
Leaf	Simple, stiff, oblong, elliptic, ovate or obovate, sometimes finely toothed

Grown up tree of Brazilian guava

Flowering in Brazilian guava

Growth form	Small tree or shrub
Flowering period	March –July, Sept Oct
Pollination	Cross
Pollinating agents	Bees
Economic yield	5 years
Yield	5-10 kg
First harvest	3 years
Harvesting method	Manually
Harvesting period	July - Oct
Physiological disorders	-
Pests	-
Diseases	-
Shelf life	2-3 days at room temperature
Shelf life in cold storage	-
Post harvest products	Jelly, juice

Fruiting branch of Brazilian guava

Mature fruits of Brazilian guava

Nutritional value of Brazilian Guava fruits (Per 100 g of edible portion)

Nutrients	Quantity	Nutrients	Quantity
Calories	74.5 kcal	Phosphorus	24.7 mg
Moisture	78 %	Iron	0.24 mg
Protein	1.8 g	Sodium	2.6 mg
Fat	1.08 g	Copper	0.18 mg
Carbohydrates	14.4 g	Magnesium	28.9 mg
Fibre	4.7 g	Manganese	0.18 mg
Calcium	15.8 mg	B- Carotene	0.65 mg
Potassium	531 mg	Lycopine	0.074 mg

7 CEYLON OLIVE

Ceylon Olive	Found in forest of Karnataka, Kerala, Maharashtra, Tamil Nadu, Manipur, Assam
Botanical name	*Elaeocarpus serratus* Linnaeus
Family	Elaeocarpaceae
Commercial varieties/lines	Land Races
Economic part	Fruit
Propagation techniques	Seed (hard to germinate, sometimes may take 2 years or more), Cuttings
Nursery period	12-15 months
Multiplication time	July - September
Spacing	8 -10 m x 8 -10 m
Climate and soil requirements	Tropical and subtropical
Leaf	Leaves simple, alternate, 5.5-12.5 cm long, 2.5-5 cm wide, elliptic, apex acuminate with blunt tip, base acute, margin serrate,
Growth form	Tree
Flowering period	Jan -Feb
Pollination	Cross pollination

Grown up tree of Ceylon olive

Flowers of Ceylon olive

Pollinating agents	Honey bees
Economic yield	After 12 years
Yield	A full grown tree produces from 800-1000 fruits
First harvest	6-8 years
Harvesting method	Manual harvesting
Harvesting period	Dec -Jan
Physiological disorders	Nil
Pests	-
Diseases	-
Shelf life	5-6 days at Room temperature
Shelf life in cold storage	-
Post harvest products	Pickle

Fruiting in Ceylon olive

Mature fruits of Ceylon olive

Nutritional value of Ceylon olive fruits (Per 100 g of edible portion)

Nutrients	Quantity	Nutrients	Quantity
Calories	28 k cal	Phosphorus	2 3 mg
Protein	0.155 g	Iron	0.4 mg
Fat	0.52 g	Calcium	5 mg
Carbohydrates	0.64 g	Thiamine	0.01 mg
Fibre	0.8 g	Vitamin C	8 mg
Ash	0.51%		

❏❏❏

8 CARAMBOLA

(STAR FRUIT)

Carambola	Grow naturaly in Karnataka, Kerala, Maharashtra, Tamil Nadu, Manipur, Assam
Botanical name	*Averrhoa carambola* L.
Family	Oxalidaceae
Commercial varieties/lines	Very sour type, richly flavored with more oxalic acid; the larger, sweet type, mild-flavored with less oxalic acid.
Economic part	Fruit
Propagation techniques	Seed, Cleft-grafting, Air-layering
Nursery period	10-12 months
Multiplication time	July -September
Spacing	6 m x 6 m

Grown up tree of carmbola

Flowering in carambola

Climate and soil requirements	Tropical and sub-tropical , thrives up to an elevation of 4,000 ft (1,200 m) evenly distributed rainfall throughout year	 **Fruiting in Carambola**
Leaf	Compound, spirally arranged, alternate leaves, 20-25 cm long, with 5 - 11 nearly opposite, ovate-oblong leaflets 3 to8 cm in length.	
Growth form	Perennial, vigorous	
Flowering period	Feb- March , July - August	
Pollination	Cross pollination	
Pollinating agents	Insects	
Economic yield	After 6 years	 **Mature fruits of carambola**
Yield	30-40 kg/tree	
First harvest	3 years	
Harvesting method	Manually	
Harvesting period	July –Sept, Nov.- Dec	
Physiological disorders	-	
Pests	fruit fly, fruit moths and fruit spotting bugs	
Diseases	Brown spot, Root rot	
Shelf life	5-7 days	
Shelf life in cold storage	2 weeks at 10°C	
Post harvest products	Pickle, Chatney	

Nutritional value of Star fruit (Per 100 g of edible portion)

Nutrients	Quantity	Nutrients	Quantity
Energy	31 Kcal	Riboflavin	0.016 mg
Carbohydrates	6.73 g	Sodium	2 mg
Protein	1.04 g	Potassium	133 mg
Total Fat	0.33 g	Calcium	3 mg
Dietary Fiber	2.80 g	Iron	0.08 mg
Vitamin A	61 IU	Magnesium	10 mg
Vitamin C	34.4 mg	Phosphorus	12 mg
Thiamin	0.014 mg	Zinc	0.12 mg

❑❑❑

9 DRAGON FRUIT (PITAYA)

Dragon fruit	Dragon fruit is originated in Central America. Fruits are rich in vitamins and minerals. Fruit has huge demand due to taste, nutritional and medicinal properties.	 Dragon fruit planning in row
Botanical name	*Hylocereus undatus* (Haw.) Britton & Rose	
Family	Cactaceae	
Commercial varieties	Red colour fruit coat with white colour flesh, Red colour fruit coat with red colour flesh , Yellow colour fruit coat with white colour flesh	
Economic part	Fruit	
Propagation techniques	Stem cutting	 Flower of Dragon fruit
Nursery period	3-5 months	
Multiplication time	June - September	
Spacing	3 x 3 m, 4 plants/pillar	
Climate and soil requirements	Tropical climate, optimum temperature range is 20 - 30° C with well distributed annual rainfall of 100-150 cm. Sandy loam soil with high organic matter and pH ranging from 5.5 to 6.5.	

Leaf	Stem green and fleshy, Leaves modify to spines	
Growth form	Fast growing climber with triangular stem, drooping	 Immature fruits of dragon fruit
Flowering period	May - August	
Pollination	Cross pollination	 Mature fruit of dragon fruit
Pollinating agents	honey bees, little honey bees and dragon flies	
Economic yield	After 3 years	
Yield/vine	15-25 kg/vine	
First harvest	18-24 months	
Harvesting method	Fruits harvested with blade or knife/secateur	 Horizontal section of Dragon fruit
Harvesting period	30 -50 days after fruit set	
Pests	Mealy bugs , Termite	
Diseases	Watery stem rot, Fruit rot	
Shelf life	One week	 Fruit rot in dragon fruit
Shelf life in cold storage	Up to three month at 8-10° C.	
Post harvest products	Fruits are used as fresh fruit. It can be used for preparation of jam, ice cream, jelly, fruit juice etc	

Nutritional value of dragon fruit (per 100g edible portion)

Nutrients	Quantity	Nutrients	Quantity
Water	82.5- 83 g	Iron	0.55- 0.65 mg
Energy	36 kcal	Carbohydrates	13.0 g
Protein	0.159- 0.229 g	Vitamin C	8.0- 9.0 mg
Fat	0.21- 0.61 g	Thiamine	0.28- 0.30 mg
Fiber	0.7- 0.9 g	Riboflavin	0.043- 0.044 mg
Carotene	0.005- 0.012 mg	Niacin	1.297- 1.30 mg
Calcium	6.3- 8.8 mg	Ash	0.28 g
Phosphorus	30.2- 36.1 mg	Other	0.54- 0.68 g

❑❑❑

10 DURIAN

Durian	Native of South Eastern Asia. Popular for strong odour and formidable thorn like pericarp and medicinal properties
Botanical name	*Durio zibethinus* L.
Family	Malvaceae
Commercial varieties	D 24, D 99 (Gob kecil), D 123 (Chanee) (In Malaysia)
Economic part	Fruit , seed
Propagation techniques	Cleft grafting
Nursery period	14-18 months
Multiplication time	June- September
Spacing	10 m x 10 m
High density	6 m x 6 m
Climate and soil requirements	Tropical climate with 200 cm rainfall, deep, well-drained loamy soils with Soil pH of 5.0-6.5. Perform well at elevation of less than 600 m msl.
Leaf	Simple, evergreen, elliptic to oblong
Growth form	Large, Upright tree

Grown up tree of durian

Flowers of durian

Immature fruits of durian

Flowering period	Jan - Feb
Pollination	Cross pollination
Pollinating agents	Bats
Economic yield	After 10 years
Yield/tree	40 - 60 kg/tree
First harvest	6 years
Harvesting method	Fruit with stalk cutting with knife or knife with long handle
Harvesting period	3-4 months after fruit set
Pests	Fruit borers, beetles, and leaf cutters
Diseases	Phytophthora root rot
Shelf life	5-7 days
Shelf life in cold storage	Up to 30 days at 4° C.
Post harvest products	Mainly used as fresh fruit or making chip, Frozen Durian, Powder, Candy

Mature fruit of durian

Horizontal section of durian fruit

Nutritional value of Durian Fruit (per 100g of edible portion)

Nutrients	Quantity	Nutrients	Quantity
Protein	1.47 g	Calcium	6.0 mg
Fat	5.33 g	Copper	0.21 mg
Total carbohydrates	27.09 g	Iron	0.43 mg
Crude fibre	3.8 g	Magnesium	30. mg
Vitamin A as carotene	44 IU	Manganese	0.032 mg
Ascorbic acid	19.7 mg	Phosphorus	39.0 mg
Niacin	1.07 mg	Potassium	436.0 mg
Riboflavin	0.2 mg	Zinc	0.28 mg
Thiamine	0.37 mg	Sodium	2.0 mg

11 EGG FRUIT

Canistel, Egg fruit	Native of southern Mexico, Belize, Guatemala and El Salvador. It is cultivated in central American countries
Botanical name	*Pouteria campechiana* Baehni
Family	Sapotaceae
Commercial varieties/lines	Ammee sapota , local races
Economic part	Fruit
Propagation techniques	Seeds , veneer grafting, cleft grafting, patch budding and air-layering
Nursery period	9-12 months
Multiplication time	July -September
Spacing	6 m x 6 m
Climate and soil requirements	Tropical or subtropical climate, moderate precipitation, deep, fertile, well-drained soil but is said to be more fruitful on shallow soil
Leaf	Evergreen, alternate but mostly grouped at the branch tips, are relatively thin, glossy, oblanceolate, or obovate, bluntly pointed at the apex, more sharply tapered at the base;

Young tree of Yellow Sapota

Flower buds of Yellow Sapota

Growth form	Perennial, vigorous
Flowering period	April - June
Pollination	Cross pollination
Pollinating agents	Wind bee
Economic yield	After 8 years
Yield/vine	100-200 fruit/tree
First harvest	9-12 months
Harvesting method	Fruit manually with or without stalk
Harvesting period	October to February
Physiological disorders	Nil
Pests	Scale insects
Diseases	Rust, fruit spot, leaf spot and scab, leaf necrosis
Shelf life	3 to 10 days
Shelf life in cold storage	-
Post harvest products	Pancakes, cupcakes, jam, and marmalade

Fruiting in Yellow sapota

Mature fruits of Yellow sapota

Nutritional value of yellow sapota fruits (per 100 g of edible portion)

Nutrients	Quantity	Nutrients	Quantity
Calories	138.8 kcal	Iron	0.92 mg
Moisture	60.6 g	Carotene	0.32 mg
Protein	1.68 g	Thiamine	0.17 mg
Fat	0.13 g	Riboflavin	0.01 mg
Carbohydrates	36.69 g	Niacin	3.72 mg
Fiber	0.10 g	Ascorbic Acid	58.1 mg
Ash	0.90 g	Tryptophan	28 mg
Calcium	26.5 mg	Methionine	13 mg
Phosphorus	37.3 mg	Lysine	84 mg

12 ELEPHANT APPLE

(KAITH)

Elephant apple, kaith	Native of India and Ceylon and is found through out India, upto an elevation of 1500 meter.
Botanical name	*Limonia acidissima* L. syn. *Ferronia elephantum* correa
Family	Rutaceae
Commercial varieties/line	There are 2 forms, one with large, sweetish fruits; one with small, acid fruits.
Economic part	Bark, Leaf, Fruit, Root
Propagation techniques	Seeds ,root cuttings, air-layers, budding
Nursery period	15 - 18 months
Multiplication time	July -September
Spacing	8 x 8 m
Climate and soil requirements	Monsoon climate with a distinct dry season, grows well in all soil types, but it is best in light soils
Leaf	Imparipinnate, Obtuse, Clustered, Cuneate, Entire, Obovate
Growth form	Tree, Upright growth
Flowering period	April -June

A grown up tree of elephant apple

Flowering Elephant apple

Fruiting in Elephant apple

Pollination	Cross pollination
Pollinating agents	Honey bees
Economic yield	After 15 years
Yield	50-100 fruits
First harvest	9-12 years
Harvesting method	Manual
Harvesting period	10-11 months after fruit set
Physiological disorders	Nil
Pests	Fruit borer
Diseases	Fruit rot in storage
Shelf life	30-40 days
Shelf life in cold storage	-
Post harvest products	Juice, Jam

Mature fruits of Elephant apple

Fungal rot in stored fruit of elephant apple

Nutritional value of elephant apple fruits (per 100g edible portion)

Nutrients	Quantity	Nutrients	Quantity
Moisture	69.5 g	Vitamin B_3	170 µg
Carbohydrate	15.5 g	Vitamin C	2.0 mg
Protein	7.3 g	Calcium	13 mg
Fat	3.7 g	Phosphorus	11 mg
Fibre	5.2 g	Magnesium	0.18 mg
Vitamin A (carotene)	61 IU	Iron	0.6 g

❒❒❒

13 GOVERNOR'S PLUM

Governor's plum	Native to Asia and Africa, A spiny shrub or small tree, up to 10 m tall
Botanical name	*Flacourtia montana* (J. Graham)
Family	Salicaceae
Commercial varieties/line	Nil
Economic part	Fruit
Propagation techniques	Seed
Nursery period	9-12 months
Multiplication time	July - September
Spacing	3 x 3 m, 4 x 4 m
Climate and soil requirements	Monsoon climate with a distinct dry season. grows in all soil types, but it is best adapted to light soils.
leaf	Simple, alternate
Growth form	Dense, compact shrub
Flowering period	Dec -Jan
Pollination	Cross pollination
Pollinating agents	Bees, Small bees
Economic yield	After 3 years

Grown uo trees of governor's plum

Flowering in Governor's plum

Yield	A full grown tree produces from 4000-1000 fruits	 Immature fruits of Governor's plum
First harvest	3- 5 years	
Harvesting method	Manually	
Harvesting period	60-70 days after fruit set	
Physiological disorders	Nil	
Pests	Nil	
Diseases	Nil	
Shelf life	1-2 days	
Shelf life in cold storage	-	 Ripe fruits of Governor's plum
Post harvest products	Squash, juice	

Nutritional value of Governor's plum fruits (per 100g edible parts)

Nutrients	Quantity	Nutrients	Quantity
Calories	94 kj	Vitamin B_3	0.4 mg
Carbohydrate	24.4 g	Vitamin C	5 mg
Protein	0.5 g	Calcium	33 mg
Fat	0.6 g	Phosphorus	17 mg
Fibre	1.2 g	Potassium	171 mg
Vitamin A (carotene)	30 IU	Iron	0.7 mg
Riboflavin	0.02 mg	Thiamine	0.01 mg

14 HOG PLUM

Hog Plum	Small deciduous tree, native to the tropical America, Naturalized in parts of tropical regions of Africa, India, Sri Lanka, Indonesia
Botanical name	*Spondias mombin* L. *Spondias pinnata*
Family	Anacardiaceae
Commercial varieties/lines	Tall with big fruits, Dwarf with small fruits
Economic part	Fruits
Propagation techniques	Seeds
Nursery period	9-12 months
Multiplication time	July -September
Spacing	6 m x 6 m
Climate and soil requirements	Tropical climate ,well adopted to arid and humid regions , diverse of soil types, but it is best adapted to light soils.
Leaf	Compound, pinnate, with 5-8 leaflets opposite pairs, oblong, acuminate
Growth form	Medium, erect, tree
Flowering period	April -June
Pollination	Cross pollination

Grown up tree of Hog plum

Flowering in Hog plum

Pollinating agents	Honey bees
Economic yield	After 8years
Yield	10-12 kg / tree
First harvest	4-5years
Harvesting method	Manually by Hand with stalk
Harvesting period	August -October
Physiological disorders	Nil
Pests	Fruit fly
Diseases	Black spot
Shelf life	3-5 days
Shelf life in cold storage	-
Post harvest products	Chutney, Pickle, Jam, Jelly

Fruiting in hog plum

Mature fruits of hog plum

Nutritional value of Hog plum fruits (per 100g edible portion)

Nutrients	Quantity	Nutrients	Quantity
Energy	48 Kcal	Vitamin B_1	0.02 mg
Moisture	90 g	Vitamin B_2	0.2 mg
Carbohydrate	4.6 g	Vitamin B_3	0.3 mg
Protein	0.7 g	Vitamin C	21 mg
Fat	0.4 g	Calcium	36 mg
Fibre	1.0 g	Phosphorus	11 mg
Vitamin A (carotene)	270 ug	Iron	3.9 mg

15 JACK FRUIT

Jack fruit	Native of India, Originated in the rain forests of the Western Ghats of India. Now spread and growing in almost whole tropical region of the world
Botanical name	*Artocarpus heterophyllus* Lam
Family	Moraceae
Commercial varieties/lines	Ghula, Hazari, Rudrakshi, Gulabi, Hazar, Singapore
Economic part	Fruits, wood
Propagation techniques	Cleft grafting, patch budding
Nursery period	12-18 months
Multiplication time	July -September
Spacing	6 m x 6 m, 8x8 m
Climate and soil requirements	Tropical or subtropical climate, deep, fertile, well-drained soil
Leaf	Evergreen, Leaves simple, alternate, spiral, lanceolate
Growth form	Perennial, vigorous
Flowering period	Feb- March and July - August
Pollination	Cross pollination

Bearing tree of Jack fruit

Flowering in Jackfruit

Pollinating agents	Small flies, gall midges	 Immature fruits of Jackruit
Economic yield	After 8 years	
Yield	50-200 fruit/tree	
First harvest	6-8 years	
Harvesting method	Fruit manually with stalk	
Harvesting period	Feb -June	
Physiological disorders	Nil	
Pests	Fruit borer, fruit flies and birds	 Carpel and seeds of Jack fruit
Diseases	Fruit rot	
Shelf life	30-40 days	
Shelf life in cold storage	Up to four weeks at 10°C temp & 75 % RH	 Fruit rot in Jackfruit
Post harvest products	Pickle, flakes, chips , powder	

Nutritional value of Ripe Jackfruit (per 100 g of edible portion)

Nutrients	Quantity	Nutrients	Quantity
Calories	95 kcal	Iron	0.23 mg
Moisture	73.5 g	Potassium	448 mg
Protein	1.72g	Sodium	2 mg
Fat	0.64 g	Zinc	0.13 mg
Carbohydrates	19.03 g	Carotene	157 µg
Fibre	1.5 g	Thiamine	0.105 mg
Magnesium	29 mg	Riboflavin	0.055 mg
Calcium	24 mg	Niacin	0.92 mg
Phosphorus	21 mg	Ascorbic Acid	13.8 mg

❑❑❑

16 JAMUN

Jamun	An evergreen tree, native to the Indian Subcontinent, spread overseas from India by Indian emigrants
Botanical name	*Syzygium cumini* (L.) *skeels*
Family	Myrtaceae
Commercial varieties/lines	Goma Priyanka, CISH J-42/CISHJ-37, Local, Gokak-1, 2, 3, Dhoopdal, Konkan Bahadoli,
Economic part	Fruit, seed, wood
Propagation techniques	Seed, cleft grafting
Nursery period	12-15 months
Multiplication time	July -September
Spacing	6 m x 6 m or 8 x 8 m
Climate and soil requirements	Tropical or subtropical climate. It requires moderate precipitation; does well in regions with a long dry season.
Leaf	Simple, alternate, glossy, oblanceolate,
Growth form	Perennial, vigorous tree
Flowering period	Feb- March
Pollination	Cross pollination
Pollinating agents	Honey bees, flies

Bearing tree of Jamun

Flowering in Jamun

Economic yield	After 8 years
Yield/tree	40-100 kg/tree
First harvest	5-6 years
Harvesting method	Manually with or with out stalk
Harvesting period	90 - 110 days after fruit set
Physiological disorders	Nil
Pests	Fruit fly/Seed borer
Diseases	Anthracnose
Shelf life	3-4 days
Shelf life in cold storage	Up to four weeks at 10° C Temp & 75 % RH
Post harvest products	Fruit juice, beverages, jellies, jam, vine etc.

Mature fruits of jamun

Fruits of Jamun

Nutritional value of Jamun fruits (per 100g edible portion)

Nutrients	Quantity	Nutrients	Quantity
Energy	60 kJ	Vitamin B_3	0.245 mg
Moisture	84.75 g	Vitamin C	11.85 mg
Carbohydrate	14 g	Calcium	11 mg
Protein	0.99 g	Phosphorus	15.6 mg
Fat	0.23 g	Potassium	55 mg
Fibre	0.6 g	Iron	1.41 mg
Thiamine	0.019 mg	Sodium	26.2 mg
Riboflavin	0.009 mg	Magnesium	55 mg

17 JANGLE JALEBI

(MANILA TAMARIND)

Jangle jalebi (Manila tamarind)	Native to the Pacific Coast and adjacent highlands of Mexico, Central America, and northern South America, Throughout India except hills
Botanical name	*Pithecellobium dulce* (Rose) Benth.
Family	Fabaceae
Commercial varieties/lines	-
Economic part	Fruit, Stem (timber wood)
Propagation techniques	Seeds
Nursery period	9-12 months
Multiplication time	July -September
Spacing	10 m x 10 m
Climate and soil requirements	Tropical climate, moderate to high precipitation, deep, fertile, well-drained soil
Leaf	Compound, Alternate-spiral, Bipinnate, Oblong-oblanceolate Leaf Base, Entire, Obtuse
Growth form	Perennial, vigorous tree
Flowering period	Feb - March

Grown up trees of Jangle jalebi

Flowering in Manila tamarind

Immature fruits of Manila tamarind

Pollination	Crops pollination
Pollinating agents	Large and Small bees
Economic yield	After 10 years
Yield	1000 - 2000 fruit/tree
First harvest	6 - 8 years
Harvesting method	Manually
Harvesting period	June -August
Physiological disorders	Nil
Pests	-
Diseases	-
Shelf life	-
Shelf life in cold storage	-
Post harvest products	-

Immature, mature and ripe fruits of manila tamarind

Nutritional value of jungle jalebi fruits (per 100g of edible portion)

Nutrients	Quantity	Nutrients	Quantity
Calories	78.8 kcal	Phosphorus	42 mg
Moisture	75.8-77.8 g	Iron	0.5 mg
Protein	2-3.3 g	Vitamin A	25 IU
Fat	0.4-0.5 g	Thiamine	0.24 mg
Carbohydrates	18.2-19.6 g	Riboflavin	0.1 mg
Fibre	1.1-1.2 g	Niacin	0.6 mg
Calcium	13 mg	Ascorbic acid	138 mg

18 KARONDA

Karonda	Karonda is an evergreen shrub or short stature tree. It grows naturally in the northern India and Western Ghats at elevations of 300 to 1800 m msl.
Botanical name	*Carissa carandas* L.
Family	Apocynaceae
Commercial varieties	Pant Manohar, Pant Sudarshan, Pant Suvarna, Konkan bold, CHES-K-II-7,CHES- K- V-6
Economic part	Fruit
Propagation techniques	Cutting, Air layering
Nursery period	12-18 months
Propagation time	March & September
Spacing	3 m x 3 m
Hedge planting	0.30 x 0.30 m
Climate and soil requirements	Grow successfully in tropical and subtropical climate, wide range of soil types, viz. sandy loams, laterite, alluvial sand, and calcareous soil even in stony , rocky and less fertile soils .but higher yield is obtained in alluvial sandy loam soils with good drainage.
Leaf	Simple

Grown up shrub of karonda

Flowering in karonda

Growth form	Upright, bushy, drooping
Flowering period	Feb- March
Pollination	Cross pollination
Pollinating agents	Honey bee
Economic yield	After 5 years
Yield/plant	5-10 kg
First harvest	3 years
Harvesting method	Fruit with stalk
Harvesting period	110-120 days after fruit set
Pests	Leaf eating caterpillar, fruit fly
Diseases	Anthracnose
Shelf life	3-4 days
Post harvest products	Pickle, Jam. Jelly, Candy, Chatney etc.

Mature fruits of Karonda

Nutritional value of karonda fruits (per 100g of edible portion)

Nutrients	Nutritional value (100 gram)	
	Fresh	**Dried**
Energy (Calorie)	42	364
Moisture (%)	91	18.2
Protein (%)	1.1	2.3
Carbohydrate (%)	2.9	67.1
Fat (%)	2.9	9.6
Mineral (%)	-	2.8
Calcium (mg)	2.1	160
Phosphorus (mg)	28	60
Iron (mg)	-	39
Vitamin C (mg)	200-500	1

19 KOKUM

Kokum	Indigenous to Western Ghats, known for anti-obesity properties	 Grown up tree of kokum
Botanical name	*Garcinia indica* Choicy	
Family	Clusiaceae	
Commercial varieties	Konkan Amruta, Konkan Hatis, CHES GI-V-8	
Economic part	Fruit , seed	
Propagation techniques	Cleft grafting, seeds	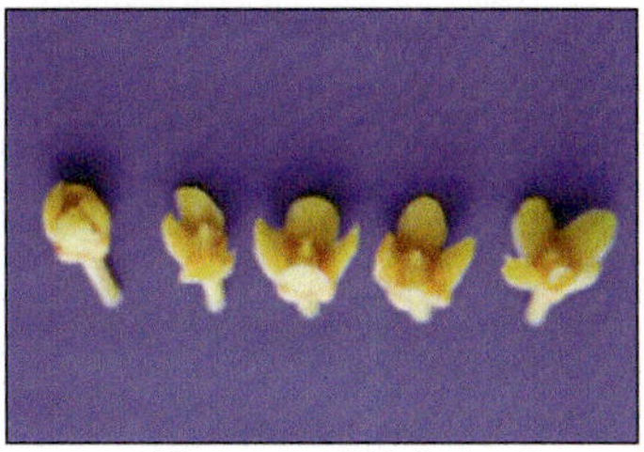 Flowers of Kokum
Nursery period	14-18 months	
Grafting time	March & September	
Spacing	6 m x 6 m	
Climate and soil requirements	Flourishes very well up to an elevation of about 800 m msl and over 250 cm rainfall. It comes up well in lateritic, alluvial soils Well drain soil.	
Leaf	Simple, dark green above & pale beneath glabourous with entire margin	 Immature fruits of kokum
Growth form	Upright, divergent tree	
Flowering period	Feb- March	
Pollination	Cross pollination	
Pollinating agents	Wind	
Economic yield	After 15 years	

Yield/tree	30-40 kg/tree
First harvest	6-8 years
Harvesting method	Individual Fruits harvested by hand plucking
Harvesting period	3-4 month after fruit set
Pests	Leaf eating beetles
Diseases	Dieing back of twigs
Shelf life	4 -5 days under ambient temperature. It can be extended
Shelf life in cold storage	Up to 28 days if stored at 13° C and 86 per cent relative humidity
Post harvest products	Kokum syrup, salted kokum syrup and dried rind, kokum butter, powder

Ripe fruits of kokum

Kokum sorbet

Nutritional value of kokum rind (moisture free basis)

Nutrients	Quantity	Nutrients	Quantity
Protein	1.92%	Starch	1%
Crude Fat (%hexane extract)	10.0%	Carbohydrates	36.4
Pectin	5.71%	Hydrocyanic acid	22.8 %
Crude fibre	14.28%	Total ash	2.57 %
Anthocyanin	2.40	Tannins	2.85%
Ascorbic acid	0.06%	-	-

❑❑❑

20 LASODA

Lasoda	Native of India, It is cultivated in central American countries
Botanical name	*Cordia dichotoma G. Frost*
Family	Boranginaceae
Commercial varieties/lines	Thar Gaurav
Economic part	Fruit
Propagation techniques	seeds
Nursery period	12 + 15 months
Multiplication time	July - September
Spacing	6 m x 6 m
Climate and soil requirements	Tropical or subtropical arid climate, perform well under semi arid and arid conditions
Leaf	Deciduous , broad, ovate, alternate and stalked with the spread

Grown up tree of Lasoda

Flowering branch of Lasoda

Growth form	Perennial, moderately vigorous
Flowering period	March -June
Pollination	Cross pollination
Pollinating agents	Insects, bee
Economic yield	After 12 years
Yield	500-4000 fruit/tree
First harvest	After 8-10 year
Harvesting method	Fruit manually with or without stalk
Harvesting period	Sept –Oct.
Physiological disorders	Nil
Pests	Fruit flies, leaf eating caterpillars
Diseases	-
Shelf life	3-5 day at room temperature
Shelf life in cold storage	-
Post harvest products	Pickle

Fruiting in Lasoda

Ripe fruits of Lasoda

Nutritional value of Lasoda fruits (Per 100 g of edible portion)

Nutrients	Quantity	Nutrients	Quantity
Calories	65 kcal	Fibre	0.3 g
Moisture	74-82.5 g	Calcium	40 mg
Protein	1.8-2.0 g	Phosphorus	60 mg
Fat	1.0 g	Iron	0.5 mg
Carbohydrates	12.2 g		

21 LONGAN

Longan	Originated from the mountain range between Myanmar and Southern China.	 Grown up tree of longan
Botanical name	*Dimocarpus longan* Lour.	
Family	Sapindaceae	
Commercial varieties	Dawuyuan, Guangyan, Chuliang, Shixia, Wuyuan Tuyan, Wulongling Chike, Dongbi, Honghezi, Gandaki Longan -1	
Economic part	Fruit	
Propagation techniques	Air layering	 Flowering in longan
Nursery period	12-18 months	
Grafting time	June -July	
Spacing	6 m x 6 m	
Climate and soil requirements	Subtropical climate, can grows well in the tropics with a short winter, annual rainfall 150 cm, Heavy, fine textured red loamy soils with high fertility & good water holding capacity.	
Leaf	Compound leave	

Growth form	Upright
Flowering period	Feb- March
Pollination	Cross pollination
Pollinating agents	Honey bees, bees
Economic yield	After 8 years
Yield/tree	30-40 kg/tree
First harvest	4- 5 years
Harvesting method	Whole panicle of fruits with a knife or scissors
Harvesting period	5 month after fruit set
Pests	Longan stink bug, mites, scales, fruit flies, aphids, stem borers, leaf eating caterpillars, flower eating caterpillars, mealy bug
Diseases	Witches' broom
Shelf life	3-4 days
Shelf life in cold storage	Up to four weeks at 4 °C.
Post harvest products	Mainly used fresh, canned longan, dried longan nuts, longan nectar and frozen longan.

Young fruits of longaan

Mature fruits of longan

Nutritional value of Longan Fruit (per 100 g of edible portion)

Nutrients	Quantity	Nutrients	Quantity
Protein	1.31 g	Calcium	1.0 mg
Fat	0.1 g	Iron	0.13 mg
Total carbohydrates	15.14 g	Magnesium	10.0 mg
Crude fibre	1.1 g	Manganese	0.052 mg
Ascorbic acid	84.0 mg	Phosphorus	21.0 mg
Niacin	0.3 mg	Potassium	266.0 mg
Riboflavin	0.14 mg	Zinc	0.05 mg
Thiamine	0.031 mg		

22 MAHUA

Mahua	Woody tree, native to India
Botanical name	Madhuca longifolia *(J. Konig) J.F. Macbr*
Family	Sapotaceae
Commercial varieties/lines	NM2, NM4, NM7 and NM8
Economic part	Flowers, fruits, seeds
Propagation techniques	Seeds, Veneer grafting
Nursery period	9-15 months
Multiplication time	July -September
Spacing	8-10 m x 8-10 m
Climate and soil requirements	Tropical and subtropical climate, can withstand drought, can grow in degraded rocky areas, salt-affected soils. However well-drained, deep loam soil are best
Leaf	Simple, crowded at apex of year's growth, narrowly oblong or obovate, rounded to acuminate at the apex, glabrous
Growth form	Perennial, vigorous tiu
Flowering period	April -June
Pollination	Cross pollination

Young tree of Mahua

Flowering in Mahua

Pollinating agents	Honey bees, Carpenter bee
Economic yield	After 15 years
Yield	60–80 kg/tree/year
First harvest	6-8 years
Harvesting method	Fruit manually with or with out stalk
Harvesting period	June-July
Physiological disorders	-
Pests	-
Diseases	-
Shelf life	1-2 days
Shelf life in cold storage	-
Post harvest products	Flowers and fruits used for preparation of distilled liquors, alcohol and portable spirits. Also used for preparation of cake, vinegar, jam, syrup and honey. Seeds oil used to manufacture of vegetable ghee, soap, greases and cosmetics.

Immature fruits of mahua

Mature fruits of mahua

Nutritional value of Mahua fruits (Per 100 g of edible portion)

Nutrients	Quantity	Nutrients	Quantity
Calories	111 kcal	Phosphorus	22 mg
Moisture	73.6 g	Iron	1.1 mg
Protein	1.37 g	Vitamin A	512 IU
Fat	1.61 g	Ascorbic acid	40.5-42.0 mg
Carbohydrates	22.69 g	-	-
Calcium	45 mg	-	-

❑❑❑

23 MALABAR TAMARIND

Malabar Tamarind	Indigenous to Western Ghats, rind of small pumpkin like fruit is rich in a substance called hydroxycitric acid. Rind powder widely used in Anti obesity formulations
Botanical name	*Garcinia gummigutta* (L.) N. Robson
Family	Clusiaceae
Commercial varieties/lines	CHES GG-IV-1
Economic part	Fruit , seed
Propagation techniques	Cleft grafting, seeds
Nursery period	14-18 months
Grafting time	June -July
Spacing	6 m x 6 m
Climate and soil requirements	Flourishes well up to an elevation of 1200 m msl and over 250 cm rainfall. It comes up well in lateritic, alluvial soils , well drain soil.
Leaf	Simple, dark green, with entire margin
Growth form	Upright tree
Flowering period	Feb- March

Grown up tree of Malabar tamarind

Flowering in Malabar tamarind

Pollination	Cross pollination	 Fruiting in Malabar tamarind
Pollinating agents	Wind	
Economic yield	After 15 years	
Yield/tree	30-40 kg/tree	 Ripe fruits of Malabar tamarind
First harvest	6-8 years	
Harvesting method	Individual Fruits harvested by hand plucking	
Harvesting period	3-4 month after fruit set	
Pests	Leaf eating beetles	
Diseases	Drying back of twigs	
Shelf life	4 -5 days under ambient temperature.	
Post harvest products	dried rind, powder, butter, seed used in anti obesity supplement	 Malabar tamarind vinegar

Nutritional value of Malabar Tamarind Rind (100 g edible portion)

Nutrients	Quantity	Nutrients	Quantity
Moisture	86.91 g	Non Reducing sugars	5.62 g
Protein	0.61 g	Fibre	3.1g
Fat	1.4 g	Potassium	169.7
Total sugars	8.6 g	Sodium	2.1
Reducing sugars	2.67 g	Tannins	1.7 g

❐❐❐

24 MANGOSTEEN

Mangosteen	Native of South Eastern Asian countries, Considered as the finest fruit of the world' or 'queen of fruits'. Exquisite flavour , only fruit in which glucose is in readily available form
Botanical name	*Garcinia mangostana* L.
Family	Clusiaceae
Commercial varieties	Jolo, large leaves and fruits of variable size , small leaves and small fruits
Economic part	Fruit and seed
Propagation techniques	Seeds, Cleft grafting
Nursery period	18-24 months
Grafting time	June -July
Spacing	8 m x 8 m
Climate and soil requirements	Frost-free regions with equatorial climate up to an altitude of 400–900 m and receiving 180-250 cm rainfall, deep, well-drained soil with high content of organic matter
Leaf	Simple, broad, leathery

Young tree of mangsteen

Mangosteen flower

Growth form	Medium-size, ever-green tree
Flowering period	Feb- March
Pollination	Parthienocarpic fruit
Pollinating agents	-
Economic yield	After 15 years
Yield/tree	40-60 kg/tree
First harvest	7-8 years
Harvesting method	Fruit with stalk
Harvesting period	4-5 month after fruit set
Physiological disorders	Gummosis and fruit splitting
Pests	Leaf eater, leaf miner, fruit borer
Diseases	Leaf blight, Diplodia fruit rot, brown root disease
Shelf life	2–3 weeks
Shelf life in cold storage	up to 50 days at 4–6 C with 85–90% relative humidity
Post harvest products	Mainly used fresh, processed products like jelly, paste, syrup and canned fruit segments

Immature fruits of Mangosteen

Mature fruits of mangosteen

Nutritional value of Mangosteen Fruit (per 100 g of edible portion)

Nutrients	Quantity	Nutrients	Quantity
Protein	0.41 g	Calcium	12.0 mg
Fat	0.58 g	Potassium	48 mg
Total carbohydrates	17.91 g	Iron	0.30 mg
Crude fibre	1.8 g	Magnesium	13.0 mg
Ascorbic acid	2.9 mg	Manganese	0.102 mg
Niacin	0.286 mg	Phosphorus	8.0 mg
Riboflavin	0.054 mg	Sodium	7.0 mg
Thiamine	0.054 mg	Zinc	0.21 mg

25 MACADAMIA NUT

Macadamia nut	Tropical nuts tree, indigenous to the eastern Australia, delicious and nutritious nut, grown in Australia, Hawaii, California, Florida and South Africa
Botanical name	*Macadamia integrifolia* F. Muell
Family	Proteaceae
Commercial varieties/lines	Kakea, Makai, Beaumont Maroochy
Economic part	Seeds (Kernels)
Propagation techniques	Seeds, wedge grafting
Nursery period	12-18 months
Multiplication time	July - September
Spacing	9-10 m x 9-10 m
Climate and soil requirements	Subtropical conditions with high summer and low winter temperature, long dry season, well-drained sandy loam soils
Leaf	Simple, Lanceolate to obovate with entire or spiny-serrated margin, arranged whorls of three to six,
Growth form	Perennial, vigorous
Flowering period	March - June

A grown up tree of Macadamia nut

Flowering in Macadamia nut

Pollination	Cross pollination
Pollinating agents	Honey bees, insects
Economic yield	After 10 years
Yield	80-90kg/tree
First harvest	3-4 years
Harvesting method	Fruit manually with stalk
Harvesting period	Oct –Nov.
Physiological disorders	Nut bases, moth, green stink bug
Pests	Anthracnose, Hunk Spot
Diseases	slow & quick decline
Shelf life	Up to 12 months
Shelf life in cold storage	-
Post harvest products	Roasted nuts are used in chocolate-coated nut, candies, bakery products and ice-creams

Mature fruits of Macadamia nut

Nuts of Macadamia

Nutritional value of Macadamia nut (100 g edible portion raw)

Nutrients	Quantity	Nutrients	Quantity
Energy	740 k cal	Vitamin	0.275 mg
Carbohydrates	13.8 g	Vitamin E	0.54 mg
Protein	7.9 g	Vitamin C	1.2 mg
Fat	75.8 g	Calcium	85 mg
Fibre	8.6 g	Phosphorus	188 mg
Thiamine	1.195 mg	Potassium	368 mg
Riboflavin	0.162 mg	Iron	3.69 mg
Niacin	2.473 mg	Zinc	1.30 mg
Pantothenic acid	0.76 mg	Magnesium	130 mg

26 MIRACLE FRUIT

Miracle fruit	Native of west Africa, Red colour fruits when eaten, causes sour foods subsequently consumed to taste sweet. This effect is due to miraculin
Botanical name	*Synsepalum dulcificum* Schumock and Thorn
Family	Sapotaceae
Commercial varieties/lines	-
Economic part	Fruit
Propagation techniques	Seeds
Nursery period	12-15 months
Multiplication time	July -September
Spacing	4m x 4 m
Climate and soil requirements	Tropical climate, free from frost and in partial shade with high humidity. grows best in soils with a pH of 4.5 to 5.8, tolerant to drought
Leaf	Leaves are 5–10 cm long, 2-3.7 cm wide and glabrous, simple

Young plant of Miracle fruit

Flowering in Miracle fruit

Growth form	Perennial, shrub of 1.0 - 4.0 m height
Flowering period	June-July and Sept- Oct.
Pollination	Cross pollination
Pollinating agents	Bees
Economic yield	After 8 years
Yield/vine	100-200 fruit/tree
First harvest	3-4 years
Harvesting method	Fruit manually with or without stalk
Harvesting period	August to December
Physiological disorders	Nil
Pests	Fruit fly
Diseases	-
Shelf life	2-3 day at room temperature
Shelf life in cold storage	-
Post harvest products	Food additive

Fruiting in Miracle fruit

Mature fruits of Miracle fruit

Nutritional value of Miracle fruit (Per 100 g of edible portion)

Nutrients	Quantity	Nutrients	Quantity
Moisture	59.55 g	Zinc	9.43 ppm
Protein	7.75 g	Iron	24.20 ppm
Fat	3.36 g	copper	6.22 ppm
Carbohydrates	18.84 g	Chromium	0.01 ppm
Fibre	4.36 g	Cobalt	0.01 ppm
Ash	4.36 g	Vitamin A	0.04 mg
Calcium	100 ppm	Ascorbic acid	22.69 mg

27 MOLSHREE

Molshree (bullet wood)	Medium-sized evergreen tree of tropical region of India, gives thick shade and fragrant flowers emit used as ornamental plant
Botanical name	*Mimusops elengi* L.
Family	Sapotaceae
Commercial varieties/lines	-
Economic part	Fruits, seeds, wood
Propagation techniques	Seeds
Nursery period	9-12 months
Multiplication time	July - September
Spacing	10 x 10 m
Climate and soil requirements	Tropical and subtropical climate , grows in all kind of soils.
Leaf	Simple, glossy, dark green, oval-shaped
Growth form	Evergreen tree reaching a height of about 16 m

Grown up tree of Molshree

Flowering in molshree

Flowering period	March -June
Pollination	Cross pollination
Pollinating agents	bees
Economic yield	After 12 years
Yield	10-15 kg /tree
First harvest	6- 8 year
Harvesting method	Manually
Harvesting period	Oct- Dec.
Physiological disorders	-
Pests	-
Diseases	-
Shelf life	3-4 days
Shelf life in cold storage	-
Post harvest products	Seed oil, it has medicinal properties pickle

Fruiting in Molshree

Mature fruits of Molshree

Nutritional value of Molshree fruits (Per 100 g of edible portion)

Nutrients	Quantity	Nutrients	Quantity
-	-	Potassium	98.54 mg
Moisture	79.27 g	Iron	0.59 mg
Protein	1.29 g	Sodium	5.16 mg
Fat	2.76 g	Fibre	1.13 g
Total sugar	15.2 g	Ascorbic acid	3.27 mg

❐❐❐

28 MONKEY JACK

Monkey Jack , Lakoocha, Ajanicakke	Native to India, specifically the sub-Himalayan region& western Ghats.
Botanical name	*Artocarpus lacucha* Buch. Ham
Family	Moraceae
Commercial varieties/lines	-
Economic part	Fruit
Propagation techniques	Seeds
Nursery period	12-15 months
Multiplication time	July -September
Spacing	6 m x 6 m
Climate and soil requirements	Tropical or subtropical climate, moderate precipitation, deep, fertile, well-drained soil
Leaf	Evergreen, alternate, glossy, lanceolate, ovate
Growth form	Perennial, vigorous
Flowering period	Feb- March
Pollination	Cross pollination
Pollinating agents	Flies, midges
Economic yield	After 15 years
Yield	100-200 fruit/tree
First harvest	8-10 years
Harvesting method	Fruit manually with stalk
Harvesting period	July -August
Physiological disorders	Nil

Grown up tree of Monkey Jack

Young fruits of Monkey jack

Immature fruits of Monkey jack

<table>
<tr><td>Pests</td><td>-</td><td rowspan="5">

Cross section of Immature fruits of Monkey jack</td></tr>
<tr><td>Diseases</td><td>-</td></tr>
<tr><td>Shelf life</td><td>-</td></tr>
<tr><td>Shelf life in cold storage</td><td>-</td></tr>
<tr><td>Post harvest products</td><td>Dry flakes, Pickle Chutney</td></tr>
</table>

Nutritional value of Money Jack fruits
(Per 100 g of edible portion)

Nutrients	Quantity	Nutrients	Quantity
Calories	66 kcal	Iron	0.5 mg
Moisture	82.1 g	Carotene	423 IU
Protein	0.7 g	Thiamine	0.02 mg
Fat	1.1 g	Riboflavin	0.15 mg
Carbohydrates	13.3 g	Niacin	0.3 mg
Fiber	2.0 g	Ascorbic Acid	135 mg
Calcium	50 mg	Phosphorus	20 mg

❑❑❑

29 NONI

Noni	Native range extends through Southeast Asia, cultivated throughout the tropics and widely naturalized.
Botanical name	*Morinda citrifolia* L.
Family	Rubiaceae
Commercial varieties/lines	No commercial varieties
Economic part	Fruits
Propagation techniques	Seed or stem cuttings
Nursery period	9-12 months
Multiplication time	July- September
Spacing	3 x 3 m/4 x 4 m
Climate and soil requirements	Tropical – subtropical with heavy rainfall to drought hit areas. grows well in all type of degraded and unfertile land soil, performs better in sandy loam soil.
Leaf	large bright green elliptical leaves
Growth form	small evergreen tree, Perennial
Flowering period	Round the year
Pollination	Self pollination
Pollinating agents	-
Economic yield	3 years onwards
Yield	40- 60 kg tree

Grown up plant of Noni

Flowering in Noni

Immature fruits of Noni

First harvest	3 years
Harvesting method	Manual picking
Harvesting period	Round the year
Physiological disorders	-
Pests	Mites, scales
Diseases	leaf and fruit blight
Shelf life	5-7 days
Shelf life in cold storage	-
Post harvest products	Juice, powders, cosmetic products, oil

Ripe fruits of Noni

Cross section of Noni

Nutritional value of Noni fruits (Per 100 g of edible portion)

Nutrients	Quantity	Nutrients	Quantity
Calories	15.3 Kcal	Dietary fibre	0.2 g
Moisture	95.67 g	Vitamin A	5 IU
Protein	0.43 g	Vitamin C	33.65 mg
Fat	0.1 g	Calcium	10.1 mg
Carbohydrates	3.40 g	Sodium	10.5 mg
Ash	0.54 g		

30 PASSION FRUIT

Passion Fruit	Passion Fruit is a native of Brazil , grown mostly in tropical and sub-tropical part of the world, Introduced in India twentieth century, fruits have unique flavor and aroma with high nutritional and medicinal properties, rich in Vitamin A, Vitamin C, sodium, magnesium, sulphur and chlorides and leaves are rich source of passiflorine, a sedative.
Botanical name	*Passiflora edulis* Sims
Family	Passifloraceae
Commercial varieties/lines	Kaveri, Ooty Purple, Coorg Purple, Moodabidri Purple, Thrissur Purple, Cherapunji Purple, Ooty Yellow, Coorg Yellow, Munnar Yellow
Economic part	Fruit, seed, leaves
Propagation techniques	Cutting, serpentine layering, Cleft grafting
Nursery period	3-5 months
Multiplication time	July - September
Spacing	2 m x 3 m

Passion fruit plantation

Flower of passion fruit

Climate and soil requirements	Tropical and sub-tropical climate with moderate rainfall (100 and 250 cm), Light to heavy sandy loam soil with a pH of 6.5 to 7.5, sensitive to water logging
Leaf	Palmate compound
Growth form	Perennial, vigorous, climbing, woody vine
Flowering period	Feb - March and July - August
Pollination	Cross pollination
Pollinating agents	Honey bees, Carpenter bee
Economic yield	After 3 years
Yield/vine	10 -12 kg/tree
First harvest	9 -12 months
Harvesting method	Fruit manually with or with out stalk
Harvesting period	60-70 days after fruit set
Physiological disorders	Nil
Pests	Mites, thrips, fruit fly
Diseases	Brown spot, Root rot, collar rot
Shelf life	3-4 days
Shelf life in cold storage	up to four weeks at 10° C Temp & 75 %RH
Post harvest products	Processed to make fruit juice and concentrate. juice delicious with unique flavour, excellent blending quality

Passion fruit vine in bearing

Fruit of Passion fruit variety Kaveri

Collar rot disease of Passion fruit
Passion fruit juice

Passion fruit Juice

Nutritional value of passion fruit (per 100 g of edible portions)

Composition	Purple passion fruit (*Passiflora edulis*)	Yellow passion fruit (*Passiflora edulis var. flavicarpa*)
Moisture (%)	85.6	84.9
Protein (g)	0.9	0.7
Fat (g)	0.1	0.2
Carbohydrate (g)	13.6	13.1
Ash (g)	0.3	0.5
Ca (mg)	3.6	3.8
P (mg)	12.5	24.56
Fe (mg)	0.2	0.4
Vitamin A (IU)	717	2410
Thiamine (mg)	Trace	Trace
Riboflavin (mg)	0.1	0.1
Niacin (mg)	1.5	2.2
Ascorbic acid (mg)	30	20

❏❏❏

31 PHALSA

Phalsa	Phalsa is a shrub or small tree of Indian origin. It has been used since Vedic period. The ripe fruits are consumed fresh or processed into refreshing fruit and soft drinks during summer.
Botanical name	*Grawia subinaequalis* DC
Family	Tiliaceae
Commercial varieties/lines	Local , Sharbati, tall Thar Pragati
Economic part	Fruit
Use of other parts	Fresh leaves used as animal fodder, mucilaginous extract of the bark used in clarifying sugar and jiggery. Fiber extracted from the bark for rope making. pruned branches used for basket making, support sticks
Propagation techniques	Cutting, air layering
Nursery period	9-12 months
Multiplication time	July-August
Spacing	3 m x 3 m

Grown up plant of Phalsa

Flowering in Phalsa

Climate and soil requirements	Subtropical fruit plant but can be grown in wide climatic conditions except high attitude (>1000m). Variable soil types including fine sand, clay or even limestone. But loamy soil is most suitable, optimum pH range- 6.1 to 6.5.	 Fruiting branch of Phlsa
Leaf	Simple	 Ripe fruits of Phalsa
Growth form	Medium, drooping shrub	
Flowering period	February-March	
pollination	Cross pollination	
Pollinating agents	Honey bees	
Economic yield	After 3 years	
Yield/tree	5-10 kg/tree	
First harvest	15 to 18 months	
Harvesting method	Hand picking, multiple harvest	
Harvesting period	40 – 55 days after fruit set	
Pests	Mites, Bark eating caterpillar, leaf eating caterpillar	
Diseases	Leaf spot, rust	 Phalsa sorbet
Shelf life	One - two days	
Post harvest products	Phalsa juice (sarbat) and Phalsa squash.	

Nutritional value of phalsa fruits (per 100 g of edible portion)

Nutrients	Quantity	Nutrients	Quantity
Moisture	80.8 g	Vitamin B_3	0.3 mg
Carbohydrate	21.1 g	Vitamin C	22 mg
Protein	1.5 g	Calcium	129 mg
Fat	0.9 g	Phosphorus	39 mg
Fibre	1.2 g	Potassium	375 mg
Vitamin A (carotene)	419 ug	Iron	3.1 mg
Energy	90 K cal		

❑❑❑

32 PUMMELO

Pummelo	Native of southern Thailand and northern Malaysia, Largest among Citrus fruits, low calorie fruit and excellent source of potassium, Thiamine, Riboflavin, niacin
Botanical name	*Citrus grandis* (L.) Osbeck.
Family	Rutaceae
Commercial varieties/Landraces	Red Flesh, white flesh, Devanahalli, Kanapara, triumph, Royal, NRC Pummelo-6, Arka Anand, Arka Chandra
Economic part	Fruits
Propagation techniques	Patch budding
Nursery period	12-18 months
Grafting time	June- July
Spacing	6 m x 6 m
Climate and soil requirements	Tropical or sub tropical climate, found growing at an altitude of 1500 m, annual rainfall - 100-150 cm, deep, medium-textured, fertile soil Intolerant to saline conditions, optimum pH range of pH - 5 to 7. Extremely sensitive to poor drainage

Fruiting tree of Pummello

Pummello flowers

Leaf	Simple
Growth form	Medium-sized tree, low spreading branches
Flowering period	June-July
pollination	Cross pollination
Pollinating agents	Honey bees
Economic yield	After 8 years
Yield/tree	50-75 kg/tree
First harvest	4- 5 years
Harvesting method	Fruit with stalk
Harvesting period	6-8 month after fruit set
Pests	Soft green scale, shoot borer, aphids , psylla, fruit fly
Diseases	Phytophthora rot, powdery mildew, Citrus Greening disease, Citrus Tristeza virus
Shelf life	Two weeks at room temperature
Shelf life in cold storage	Up to 3 weeks at 12 C,
Post harvest products	Mainly used fresh, fruit chats

Fruiting in Pummelo

Mature fruits of Pummello

Phytophthora rot

Nutritional value of pummelo Fruit (per 100 g of edible portion)

Nutrients	Quantity	Nutrients	Quantity
Energy (Calorie)	25-58	Phosphorus	20-27 mg
Moisture	82-94 g	Iron	0.1-3.0 mg
Protein	0.7-2.0 g	Potassium	150-225 mg
Carbohydrate	3-12 g	Vitamin A	20-30 mg
Fat	0.05-0.5 g	Vitamin B	0.2-.04 mg
Calcium	21-30 mg	Vitamin C	30-43 mg

❑❑❑

33 PULASAN

Pulasan	Attractive evergreen tree, native to Western Close relative of rambutan
Botanical name	*Nephelium mutabile* (Labill) Leenh
Family	Sapindaceae
Commercial varieties/lines	Jolo, Seebabat, Kapoolasan seebabat
Economic part	Fruits
Propagation techniques	Seeds, recalcitrant, can not be stored for longer periods, patch or shield Budding
Nursery period	12-18 months
Multiplication time	July -September
Spacing	8-10 m x 8-10 m
Climate and soil requirements	Orthodox-tropical and thrives only in very hot humid regions between 110 - 350 m of altitude, well - drained, deep loam soil are best
Leaf	Compound, pinnate or odd-pinnate, 17-45 cm long, leaflets, oblong - or elliptic-lanceolate, 6.25 -17.5 cm long and up to 5 cm wide; slightly wavy, dark-green colour
Growth form	Perennial, vigorous Tree

Mature fruit of Pulasan

Flowering in Pulasan

Flowering period	March -April
Pollination	Cross pollination
Pollinating agents	Bees, Flies
Economic yield	After 15 years
Yield	20-40kg/tree
First harvest	6-8 years
Harvesting method	Fruit manually without stalk
Harvesting period	July-August
Physiological disorders	-
Pests	Aphids
Diseases	Corky bark disease
Shelf life	Upto on week
Shelf life in cold storage	-
Post harvest products	Fresh fruits, Jam, Jelly

Immature fruits of Pulasan

Cross section of Fruits

Nutritional value of Pulasan fruits (Per 100 g of edible portion)

Nutrients	Quantity	Nutrients	Quantity
Moisture	87.62 g	Vitamin B_3	-
Carbohydrate	12.86 g	Vitamin C	11 mg
Protein	0.82 g	Calcium	0.03 mg
Fat	0.55 g	Phosphorus	-
Fibre	0.14 g	Potassium	-
Ash	0.44 g	Iron	0.002 mg

34 RAMBUTAN

Rambutan	Indigenous to the Malay Archipelago
Botanical name	*Nephelium lappaceum* var. lappaceum L.
Family	Sapindaceae
Commercial varieties	Arka Coorg Arun, Arka Coorg Peetabh, N-18, N-35
Economic part	Fruit (Aril)
Propagation techniques	Cleft grafting, Chip budding, Approach grafting
Nursery period	14-18 months
Grafting time	March & September
Spacing	8 m x 8 m
High density	6 m x 6 m
Climate and soil requirements	Grows well up to 700 metre above sea-level. requires 22-30 C temperature for optimum growth, very sensitive to low temperature (< 10o C), rainfall of 200 - 500 cm well distributed sandy loam to clay loam soils with organic matter, optimum soil pH is 4.5 to 6.5, sensitive to water logging
Leaf	Alternate, 10 – 30 cm long, pinnately Compound, with 3 -11 leaflets, each leaflet 5 – 15 cm long and 3 -10 cm broad, with an entire margin.

A grown up tree of Rambutan

Flowering in Rambutan

Growth form	Upright, divergent, drooping tree
Flowering period	Feb- May
pollination	Cross pollination
Pollinating agents	Bees , Flies
Economic yield	After 8 years
Yield/tree	30-40 kg/tree
First harvest	4- 5 years
Harvesting method	Fruits in branches along with leaves
Harvesting period	4-5 month after fruit set
Physiological disorders	Leaf scorching, fruit cracking
Pests	Mealy bugs, Beetles, Leaf eating caterpillar
Diseases	Fruit rot, Twig blight, Powdery mildew
Shelf life	3- 5 days
Shelf life in cold storage	up to two weeks at 10 -12 C.
Post harvest products	Mainly used fresh, can be used for making jams and jellies, or canned.

Fruiting branch of Rambutan

Fruits of Rambutan

Nutritional value of Rambutan Fruit (per 100 g of edible portion)

Nutrients	Quantity	Nutrients	Quantity
Energy	82 K cal	Moisture	78.3 g
Protein	0.65 g	Thiamine	0.013 mg
Fat	0.21 g	Iron	0.35 mg
Carbohydrates	20.87 g	Magnesium	7 mg
Fibre	0.9 g	Manganese	0.343 mg
Ascorbic acid	4.9 mg	Phosphorus	9.0 mg
Niacin	1.352 mg	Sodium	11.0 mg
Riboflavin	0.022 mg	Calcium	22 mg

35 ROSE APPLE

Rose apple	Native of India, East Indies and Malaya, It is cultivated in Topical Asias central American countries
Botanical name	*Syzyygium jambos* L.
Family	Myrataceae
Commercial varieties/lines	-
Economic part	Fruit
Propagation techniques	Seeds , cutting, veneer grafting and air-layering
Nursery period	12-15 months
Multiplication time	July -September
Spacing	6 m x 6 m
Climate and soil requirements	Tropical or subtropical climate, moderate precipitation, deep, fertile, well-drained soil but is said to be more fruitful on shallow soil
Leaf	Simple, evergreen leaves are opposite, lanceolate or narrow-elliptic, tapering
Growth form	Perennial tree reaching 7.0-12 m in height
Flowering period	Feb- March
Pollination	Cross pollination
Pollinating agents	Honey bees

Grown up rose apple tree

Flowers of Rose apple

Economic yield	After 12 years
Yield	100-200 fruit/tree
First harvest	After 6-7 years
Harvesting method	Fruit manually with or without stalk
Harvesting period	May -August
Physiological disorders	Nil
Pests	Aphids
Diseases	Leaf spot, root rot , Anthracnose
Shelf life	1-2 days
Shelf life in cold storage	-
Post harvest products	Fresh, jam, jelly

Fruiting in Rose apple

Mature Fruits of Rose apple

Ripe Fruits of Rose apple

Nutritional value of Rose apple fruits (Per 100 g of edible portion)

Nutrients	Quantity	Nutrients	Quantity
Energy	56 kcal	Phosphorus	11.7-30 mg
Moisture	84.5-89.1 g	Iron	0.45-1.2 mg
Protein	0.5-0.8 g	Vitamin A	123-235 IU
Fat	0.2-.03 g	Thiamine	0.01-0.19 mg
Carbohydrates	9.7-14.2 g	Riboflavin	0.03-0.05 mg
Fibre	0.7-1.9 g	Niacin	0.3 mg
Calcium	10-45.2 mg	Ascorbic acid	3-37 mg

36 SOUR SOP

<table>
<tr><td>Sour Sop</td><td>Native of South America, found both wild and cultivated from southern Mexico to Peru and Argentina up to an altitude of 1100 m msl. In India, it was introduced in the eighteenth and nineteenth centuries</td><td rowspan="5">
Young tree of Soursop</td></tr>
<tr><td>Botanical name</td><td>Annona muricata L.</td></tr>
<tr><td>Family</td><td>Annonaceae</td></tr>
<tr><td>Commercial varieties/lines</td><td>Sweet sour sop, very sour Sop</td></tr>
<tr><td>Economic part</td><td>Fruit and leaves</td></tr>
<tr><td>Propagation techniques</td><td>Seedlings,</td><td rowspan="7">
Flower of Soursop</td></tr>
<tr><td>Nursery period</td><td>9-12 months</td></tr>
<tr><td>Multiplication time</td><td>June-July</td></tr>
<tr><td>Spacing</td><td>6 m x 6 m, 5 × 5m</td></tr>
<tr><td>Climate and soil requirements</td><td>Tropical humid climate, very sensitive to low temperature, sandy to sandy loam soils of medium texture are most suitable. Soil pH - 6.0 and 6.5</td></tr>
<tr><td>Leaf</td><td>Simple</td></tr>
</table>

Growth form	Upright, divergent tree
Flowering period	May-June
Pollination	Cross pollination
Pollinating agents	Beetles, Ants
Economic yield	After 8 years
Yield/tree	30-40 kg/tree
First harvest	3-4 years
Harvesting method	Fruit with stalk
Harvesting period	2-3 month after fruit set
Pests	Mealy bugs , fruit fly, cerconota moth
Diseases	Anthracnose, Black canker, diplodia rot, Marginal leaf scorch
Shelf life	One –two days
Post harvest products	Mainly used fresh, squash, Puree and juice concentrates, the pulp preserved by freezing.

Immature fruits of sour sop

Verticla section of ripe fruit

Seeds of sour sop

Nutritional value of Sour sop Fruit (per 100 g of edible portion)

Nutrients	Quantity	Nutrients	Quantity
Energy	66 Kcal	Moisture	81 g
Protein	1.0 g	Calcium	14 mg
Fat	0.3 g	Iron	0.6 mg
Total carbohydrates	16.84 g	Magnesium	21 mg
Crude fibre	3.3 g	Potassium	278 mg
Ascorbic acid	20.6 mg	Phosphorus	27 mg
Niacin	0.9 mg	Sodium	14 mg
Riboflavin	0.05 mg	Sulphur	0.1 mg
Thiamine	0.07 mg		

37 STRAWBERRY GUAVA OR RED CATTLEY GUAVA

Strawberry guava / Red Cattley Guava	A slow growing shrub or small tree between 2 and 4m. Attractive foliage and fruit than the common guava, native to the lowlands of eastern Brazil
Botanical name	*Psidium cattleianum* Sabine
Family	Myrtaceae
Commercial varieties/lines	-
Economic part	Fruits
Propagation techniques	Seed, layering,
Nursery period	12 months
Multiplication time	July - Sept
Spacing	3 × 3 m

Grown up plant of Strawberry guava

Flowers of Strawberry guava

Climate and soil requirements	Tropical and subtropical more cold hardier than the common guava does well in limestone and poor soils, drought tolerant
Leaf	Simple, alternate, evergreen, obovate, dark, smooth, glossy, somewhat leathery, bigger than guava
Growth form	Small tree or shrub
Flowering period	March –July, Sept Oct
Pollination	Cross pollination
Pollinating agents	Bees
Economic yield	5 years
Yield	5-10 kg
First harvest	3 years
Harvesting method	Manually
Harvesting period	July -Oct
Physiological disorders	-
Pests	-
Diseases	-
Shelf life	3 to 4 days at room temperature
Shelf life in cold storage	-
Post harvest products	-

Immature fruits of Strawberry guava

Ripe fruits of Strawberry guava

Cross section of Strawberry guava fruits

Nutritional value of Red Cattley Guava fruits
(Per 100 g of edible portion)

Nutrients	Quantity	Nutrients	Quantity
Calories	38 kcal	Phosphorus	20 mg
Moisture	81.7-85.0 g	Iron	1.2 mg
Protein	0.75-1.03 g		
Fat	0.15-0.2 0g	Thiamine	0.02 mg
Carbohydrates	6.14 g	Riboflavin	0.02 mg
Fibre	9 g	Niacin	0.3 mg
Calcium	50 mg	Ascorbic acid	15-20 mg

❐❐❐

38 SURINAM CHERRY

Surinam cherry	Native to tropical South America, ranging from Surinam, French Guiana to southern Brazil, as well as parts of Paraguay, Argentina and Uruguay	 Grown up plant of Surinam cherry
Botanical name	*Eugenia uniflora* L.	
Family	Myrtaceae	
Commercial varieties/lines	Acerola, Florida sweet	
Economic part	Fruits	
Propagation techniques	It can be propagated by seeds or cutting.	
Nursery period	12-15 months	
Multiplication time	July –September	 Flowering in Surinam cherry
Spacing	3 m x 3m	
Climate and soil requirements	Tropical climate, cannot tolerate lower than -1 C. Sandy, It prefers dry, sandy, Well-drained soils.	
Leaf	Evergreen, simple	
Growth form	Perennial, shrub	
Flowering period	Feb- March and July – August	
Pollination	Cross pollination	

Pollinating agents	Bees, Flies
Economic yield	After 5 years
Yield	100-200 fruit/tree
First harvest	After 3 years
Harvesting method	Fruit manually with or without stalk
Harvesting period	June – Oct.
Physiological disorders	Nil
Pests	-
Diseases	-
Shelf life	-
Shelf life in cold storage	-
Post harvest products	Pickle, Fresh fruit

Fruiting in Surinam cherry

Ripe fruits of Surinam cherry

Nutritional value of Surinam cherry fruits
(Per 100 g of edible portion)

Nutrients	Quantity	Nutrients	Quantity
Calories	43-51 kcal	Phosphorus	11 mg
Moisture	90.7 g	Iron	0.2 mg
Protein	0.84-1.01 g	Vitamin A	1200-2000 IU
Fat	0.4-0.88 g	Thiamine	0.03 mg
Carbohydrates	7.93-12.5 g	Riboflavin	0.04 mg
Fibre	0.34-0.6 g	Niacin	0.03 mg
Calcium	9 mg	Ascorbic acid	20-30 mg

39 TAMARIND

Tamarind	Native of Tropical Africa, introduced and naturally grown in India and other parts of tropics
Botanical name	*Tamarindus indica* L.
Family	Fabaceae
Commercial varieties/lines	PKM 1, Urigam, Prathisthan, DTS 1, Yogeshwari
Economic part	Fruit
Propagation techniques	Seeds , cleft grafting, soft wood
Nursery period	9-12 months
Multiplication time	July -September
Spacing	8-10m x 8-10m, 6 × 6m
Climate and soil requirements	Tropical or subtropical climate up to1000 m msl moderate precipitation, Grown on variety of soils ranging from poor degraded, eroded, gravelly, saline and alkaline soils.
Leaf	Compound, Paripinnate
Growth form	Perennial, vigorous tree
Flowering period	April -June

Young plant of Tamarind

Flowering branch of Tamarind

Pollination	Cross pollination	 Fruiting in Tamarind
Pollinating agents	Honey bees,	
Economic yield	After 10-15 years	
Yield/vine	150 - 200 kg of fruits/ tree/year.	
First harvest	After 8-10 years	
Harvesting method	Manually with stick	
Harvesting period	9-11 months after fruit set	
Physiological disorders	Nil	 Mature fruits of Tamarind
Pests	Scale, Mealy, Bug, Pod borer, seed borer	
Diseases	Powery mildew, Leaf Spot	
Shelf life	2-6 months	
Shelf life in cold storage	-	
Post harvest products	Powder, blocks,	

Nutritional value of tamarind fruits (Per 100 g of edible portion)

Nutrients	Quantity	Nutrients	Quantity
Energy	kcal	Phosphorus	34-78 mg
Moisture	17.8-35.8 g	Iron	0.2-0.9 mg
Protein	2-3 g	-	-
Fat	0.6g	Thiamine	0.33 mg
Carbohydrates	41.1-61.4 g	Riboflavin	0.1 mg
Fibre	2.9 g	Niacin	1.0 mg
Calcium	34-94 mg	Ascorbic acid	44 mg

40 WAX APPLE

Wax Apple	Native of Malay Peninsula, Indonesia, Andaman and Nicobar Islands, widely cultivated in the tropical region,
Botanical name	*Syzygium samarangense*
Family	Myrtaceae
Commercial varieties/lines	CHESM-1, Local
Economic part	Fruit
Propagation techniques	Stem cutting, air layering
Nursery period	10-12 months
Grafting time	July- September
Spacing	6 m x 6 m
Climate and soil requirements	Tropical climate with good rainfall,
Leaf	Simple
Growth form	Upright, divergent, dense
Flowering period	Feb- March
Pollination	Cross pollination
Pollinating agents	Honey bees
Economic yield	After 8 years
Yield/tree	60-70 kg/tree

Grown up tree of Wax apple

Flower of Wax apple

First harvest	4 years
Harvesting method	Fruit with stalk
Harvesting period	2 -3 month after fruit set
Physiological disorders	Nil
Pests	Fruit Fly
Diseases	Root Rot
Shelf life	3-7 days
Shelf life in cold storage	
Post harvest products	Mainly used fresh, Juice

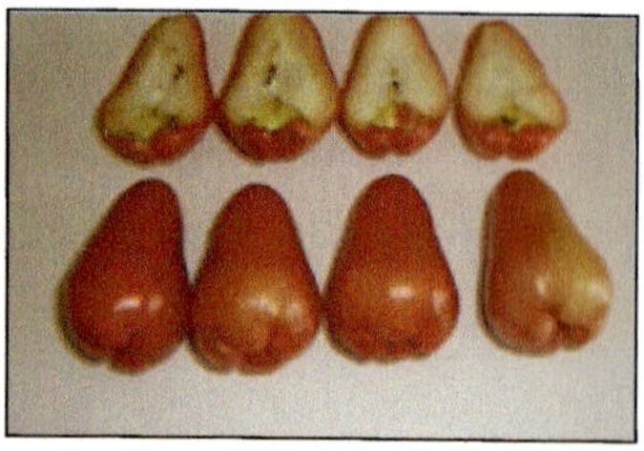

Section of Wax apple fruits

Nutritional value of Wax apple Fruit (per 100 g of edible portion)

Nutrients	Quantity	Nutrients	Quantity
Protein	0.6 g	Calcium	29 mg
Fat	0.3 g	Iron	1.0 mg
Total carbohydrates	5.7 g	Magnesium	5.0 mg
Ascorbic acid	22.30 mg	Phosphorus	8.0 mg
Niacin	0.8 mg	potassium	123 mg
Riboflavin	0.03 mg	Sulphur	0.06 mg
Thiamine	0.02 mg	Zinc	0.06 mg
Energy	25 Kcal	Moisture	93 g

❒❒❒

41 WILD JACK FRUIT

Wild Jackfruit	Native to India, specifically Western Ghats.
Botanical name	*Artocarpus hirsutus* Lam.
Family	Moraceae
Commercial varieties/lines	-
Economic part	Fruit, Stem (timber wood)
Propagation techniques	Seeds
Nursery period	9-12 months
Multiplication time	July - September
Spacing	10 m x 10 m
Climate and soil requirements	Tropical climate, moderate to high precipitation, deep, fertile, well-drained soil
Leaf	Simple, alternate, Evergreen, alternate, glossy, ovate, obovate
Growth form	Perennial, vigorous tree
Flowering period	Jan - March
Pollination	Cross pollination

Grown up tree of Wild jack

Leaf and young fruits of Wild jack

Pollinating agents	Small Flies, Midges
Economic yield	After 15 years
Yield	100-200 fruit/tree
First harvest	8 -10 years
Harvesting method	Fruit manually with stalk
Harvesting period	June - August
Physiological disorders	Nil
Pests	-
Diseases	Pink disease, Macrophomina leaf spot
Shelf life	-
Shelf life in cold storage	-
Post harvest products	-

Fruits of Wild jack

Section of fruits of Wild jack

Nutritional value of wild jack fruits (Per 100 g of edible portion)

Nutrients	Quantity	Nutrients	Quantity
Moisture	72.83 g	Carotene	3.63 µg/100 g
Carbohydrates	10.78 g	Ascorbic Acid	3.96 mg
Fiber	1.85 g	Calcium	11.53 mg
TSS	20.4^{0} Brix	Phosphorus	1.33 mg
Acidity	0.84%	Potassium	341.42 mg
		Iron	1.3 mg

42 YELLOW MANGOSTEEN

Yellow mangosteen	Native to western Ghats and North eastern India, A slander tree, up to 15 m tall
Botanical name	*Garcinia xanthochymus* Hook F.
Family	Clusiaceae
Commercial varieties/lines	Nil
Economic part	Fruits
Propagation techniques	Seeds
Nursery period	12-18 months
Multiplication time	July -September
Spacing	6 x 6 m
Climate and soil requirements	Tropical humid climate. Grows in all soil types, lateritic, alluvial soils Well drain soil..
Leaf	Simple
Growth form	Dense compact tree
Flowering period	Feb- June
Pollination	Cross pollination
Pollinating agents	Small Bees, Ant, Wind
Economic yield	After 10 years
Yield	A full grown tree produces from 100-200 fruits
First harvest	6-8 years
Harvesting method	Manually
Harvesting period	100-120 days after fruit set

Young tree of Yellow mangosteen

Flowers of Yellow mangosteen

Immature fruits of Yellow mangosteen

Physiological disorders	Nil
Pests	Nil
Diseases	Fruit Rot
Shelf life	1-2 days
Shelf life in cold storage	-
Post harvest products	Squash, dried flakes

Mature fruits of Yellow mangosteen

Nutritional value of Yellow mangosteen fruits (per 100 g edible portion)

Nutrients	Quantity	Nutrients	Quantity
Moisture	80.96 %	TSS	12.3° Brix
Calories	96.48 kj	Acidity	4.88 %
Carbohydrate	11.44%	Vitamin C	11.62 mg
Protein	5.01 %	Calcium	1.39%*
Ash	1.65 %	Potassium	0.5 %*
Fat	1.65 %	Manganese	2.9 ppm*
Fibre	2.73 %	Zinc	9.2 ppm*
Careniods	33.3 ug	Iron	56 ppm*
Anthocyanin	1.62 mg	Copper	8.4 ppm*

*on dry weight basis

❑❑❑

Selected References :

1. Ali Esmail and Al-Snafi 2016. The Pharmacological and therapeutic importance of Cordia myxa- A review. *IOSR Journal of Pharmacy*. 6(6): 47-57.
2. Arora, R.K. 1985. Genetic Resources of Less Known Cultivated Food Plants, New Delhi, India: National Bureau of Plant Genetic Resources. NBPGR Sci. Mongr. No. 9.
3. Arora, R. K. 1997. Genetic resources of native tropical fruits in Asia: Diversity, distribution and IPGRI's emphasis on their conservation and use. In: Tropical Fruits in Asia: Diversity, Maintenance, Conservation and Use. Proc. IPGRI-ICAR-UTFANET Regional Training Course on the Conservation and Use of Germplasm of Tropical Fruits in Asia Edited by: Arora, R. K. and Ramanatha Rao, V. Bangalore, India 18–31 May.
4. Collins, W.W. and Hawtin, G.C. 1999. Conserving and using crop plant biodiversity in agroecosystems. In: WW Collins, CO Qualset, eds. Biodiversity in agroecosystems. Boca Raton, Washington: CRC Press. 267–281.
5. Dahiru, D., Sini, J.M. and John Africa, L. 2006. Antidiarrhoeal activity of Ziziphus maurtiana root extract in rodents. *African Journal of Biotechnol.* 5:10.
6. FAO. 2010. The Second Report on the State of the World's Plant Genetic Resources for Food and Agriculture. Commission on Genetic Resources for Food and Agriculture. Food and Agriculture Organization of the United Nations, Rome, Italy.
7. Frison, E.A., Cherfas, J.and Hodgkin, T. 2011. Agricultural biodiversity is essential for a sustainable improvement in food and nutrition security. Sustainability 3:238–253.
8. Giuliani, A. 2007. Developing markets for agrobiodiversity: securing livelihoods in dryland areas. Earthscan Research Editions, London.
9. Gopalan, C., Ramashastri, B. V., Balasubramanian, S. C., Rao, N. B. S., Deosthale, Y. G. and Pant,K. C. 1989. Nutritive value of Indian Foods, ICMR Hyderabad: National Institute of Nutrition.
10. Gopalan, C., Ramashastri, B. V., Balasubramanian, S. C., Rao, N. B. S., Deosthale, Y. G. and Pant,K. C. 1994. Nutritive Value of Indian Foods, ICMR Hyderabad: National Institute of Nutrition.
11. Gopalan, C., Ramashastri, B. V., Balasubramanian, S. C., Rao, N. B. S., Deosthale, Y. G. and Pant, K. C. 2004. Nutritive Value of Indian Foods, ICMR Hyderabad: National Institute of Nutrition.

12. Kader, A. A., Veazie, P. P. and Lester, G. E. 2005. Nutritional Quality of Fruits, Nuts, and Vegetables and Their Importance in Human Health, Department of Pomology, University of California, Davis, CA.
13. Kulkarni, D. K. 2007. Germplasm collection of High value wild fruit species for global nutraceutical. In: Proceeding of the National Workshop on `Underutilised Fruit Species for Food–Nutrition Security and Enhanced Rural Livelihood' held on 14-17 Sep. 2005 at BAIF, Development Research Foundation, Pune. (Eds. Daniel, J.N., Hegde, N.G. and Dhar S.): 42-46.
14. Lutaladio, N., Burlingame, B. and Crews, J. 2010. Horticulture, biodiversity and nutrition. *Journal of Food Composition and Analysis*, 23: 481–485.
15. Nile, Shivraj Hariram. 2014.Edible berries: Bioactive components and their effect on human health. *Nutrition* 30 : 134–144.
16. Pareek, O. P. , Sharma, S. and Arora, R.K. 1998. Underutilised edible fruits and nuts-An inventory of genetic resources of their regions of diversity. IBPGR office of South Asia. New Delhi, India, 235p.
17. Pareek, O. P. and Sharma, S. 1993. Underutilised fruits. *Indian Horticulture*, 38(1): 48-49.
18. Quebedeaux, B. and Eisa, H. M. 1990. Horticulture and human health. Contributions of fruits and vegetables. Hort. and Human Health. Hort Science, 25: 1473–1532. Proc. 2nd Intl. Symp.
19. Sakthivel, T., Karunakaran, G., Tripathi, P. C. and Chithraichlevan, R. 2012. Status of Rambutan germplasm (*Nephelium lappaccum* L.) at CHES, Chettalli, Souvenir, Rambutan Seminar Nov. 23 Central Horticultural Experiment Station (IIHR), Chettalli, pp. 2-6.
20. Sankar, V., Tripathi, P.C., Karunakaran, G. and Senthil Kumar, R. 2014. Pummelo, Technical Bulletin8/2014, ICAR-IIHR, Central Horticultural Experiment Station, Chettalli, Kodagu, Karnataka. pp. 10.
21. Senthil Kumar, R., Tripathi, P. C., Sankar, V. and Karunakaran, G. 2014. Kokum - A potential fruits of Western Ghats. Technical Bulletin 6-/2014, ICAR-IIHR, Central Horticultural Experiment Station, Chettalli, Kodagu, Karnataka pp.10.
22. Senthilkumar, R., Sankar,V. and Tripathi, P.C. 2015. Souvenir and Abstract of workshop on Garcinia . CHES Chettalli August 20. pp.119.
23. Shamina Azeez, Karunakaran, G., Shivashankara, K S , Keshava Rao, V. Tiwari, R. B., Sathisha, G. C., Roy, T. K., Tripathi, P. C., Bharathi, L. K. and Narayana, C K. 2019. Nutritional characterstics of underutilized crops with future potential, IIHR Tech Bull. No. 03/2019. pp. 31.

24. Singh, H.P., V.A. Parthasarathy, V.A. and Prasath, D. 2009. Horticultural Crops – Varietal wealth. Studium Press (India) Pvt Ltd, New Delhi, 450p.

25. Singh, S. R., Phurailatpam, A.K., Wangchu, L., Ngangbam, P. and Chanu, T.M. 2014. Traditional medicinal knowledge of underutilized minor fruits as medicine in Manipur. *International Journal of Agricultural Sciences* 4(8): 241-247.

26. Tiwari, R.B., Sarojini Jalali and Tripathi, P.C. 2014. Dehydrated processed products from underutilized fruits work done at IIHR. Paper presented in National seminar on strategies for conservation, improvement and utilization of underutilized fruits. December 1-3, 2014., pp 365.

27. Tripathi, P. C. 2018. Passion Fruit.(In)Peter, K.V. (Ed). Horticultural Crops of high nutraceutical values. Brillion Publishing New Delhi p. 245-270

28. Tripathi, P.C., Sankar, V. Senthilkumar R. and Priti Sonavane. 2015. Souvenir and Abstract of Brain Storming session on avocado 2015. CHES, Chettalli pp. 67.

29. Tripathi, P.C., Karunakaran, G. and Sakthivel, T. 2011. Promising lines of underutilized fruits at CHES, Chettalli, pp 8.

30. Tripathi, P. C., Karunakaran, G., Sankar V. and Kanupriya. 2019. Genetic Resources of wild and underutilized fruits in Western Ghats In Shaping Future of Indian Horticulture (Ed Chadha et al). Astral International, New Delhi. pp. 143-158.

31. Tripathi, P. C., Karunakaran, G., Sankar, V. and Senthil Kumar, R. 2014. Karonda - A Potential fresh fruit of future. Technical Bulletin 7/2014, ICAR-IIHR, Central Horticultural Experiment Station, Chettalli, Kodagu, Karnataka. pp. 14.

32. Tripathi, P. C., Karunakaran ,G., Sankar, V. and Senthil Kumar, R. 2014. Dragon fruit – Nutritive and Ruminative fruit. Technical Bulletin 11/2014, ICAR-IIHR, Central Horticultural Experiment Station, Chettalli, Kodagu, Karnataka. pp.10.

33. Tripathi, P. C., Sankar, V. and Senthil Kumar, R. 2014 Sour Sop – An emerging fruit of future. Technical Bulletin, 5/2014, ICAR-IIHR, Central Horticultural Experiment Station, Chettalli, Kodagu, Karnataka.PP14.

34. Tripathi, P. C. 2018. Minor fruits for Nutritional security. In: Zero Hunger India Ed. Dr. K. V. Peter, Brillion Publishing, New Delhi pp. 706.

35. Tripathi, P.C. , Karunakaran, G., Sakthivel, T. , Sankar, V. and Senthil Kumar, R. 2014. Passion fruit cultivation in India. Technical Bulletin 3/2014, ICAR-IIHR, Central Horticultural Experiment Station, Chettalli, Kodagu, Karnataka. pp. 18.

36. Tripathi, P.C., Karunakaran, G., Sakthivel, T. Sankar, V. and Senthil Kumar 2014. Avocado cultivation in India. Technical Bulletin. 2/2014, ICAR-IIHR, Central Horticultural Experiment Station, Chettalli, Kodagu, Karnataka, Kodagu, Karnataka 18p.

37. Tripathi, P.C., Karunakaran, G., Sakthivel, T. V. Sankar, V. and R. Senthil Kumar 2014. Rambutan cultivation in India. Technical Bulletin 1/2014, ICAR-IIHR, Central Horticultural Experiment Station, Chettalli, Kodagu, Karnataka, Kodagu, Karnataka pp. 18.

38. Tripathi, P.C., Sankar V., Senthilkumar R. and Karunakaran G. 2014. Souvenir and Abstracts of National seminar on strategies for conservation, improvement and utilization of underutilized fruits. CHES, Chettalli pp. 419.

39. Tripathi, P.C. 2018. Minor fruit crops of humid tropics. In Books of abstract of International Symposium on Orchids, strawberry, minor fruits and flowers, RARS, Ambalvayal (KAU), Kerala, Jan 13-15, 2018, pp. 12-13.

40. Tripathi, P.C. Karunakaran, G. Sankar, V. and Senthilkumar, R. 2017. Diversity and utilization of future fruit crops of humid region. Souvenir And Abstracts. National Conference on Horticultural Crops of Humid Tropics - Diversification for Sustainability" at Madikeri May 20-21, pp. 20-26.

41. Tripathi, P.C., Karunkaran, G., Sankar, V. and Senthil Kumar, R., 2015. Survey and Conservation of Indigenous Fruits of Western Ghats., *Journal of Agricultural Science and Technology* A (5) : 608-615.

❑❑❑

Index